MIGRANTS IN THE DIGITAL PERIPHERY

MIGRANTS IN THE DIGITAL PERIPHERY

New Urban Frontiers of Control

MATT MAHMOUDI

UNIVERSITY OF CALIFORNIA PRESS

University of California Press
Oakland, California

Cataloging-in-Publication data is on file at the Library of Congress.

ISBN 978-0-520-39700-2 (cloth)

ISBN 978-0-520-39701-9 (pbk.)

ISBN 978-0-520-39702-6 (ebook)

33 32 31 30 29 28 27 26 25 24
10 9 8 7 6 5 4 3 2 1

To my beautiful family

Contents

Acknowledgments

THIS BOOK IS THE CULMINATION of a series of intellectual curiosities, personal struggles, and, most of all, an imperative to intervene in developments that stand to harm communities of whom I am a product. Migrant communities far and wide, communities of color, in particular fugitives from oppressive regimes who yet again face surveillance, exploitation, and marginalization, nowadays at the hands of an ever-powerful technology industry. Communities of audacity, whose political imagination is a threat against a depraved status quo. This work is in honor of them—in particular, communities on the move, delivered into the violence and racial gag order of Fortress Europe and the state-sanctioned long deaths commissioned by the United States. In this spirit, I have sought to uncover the everyday structures and practices that generate vulnerability, precarity, and exploitation, with this work.

I cannot express enough gratitude to my informants, collaborators, and research partners in Berlin, New York, and places in between, who generously shared their time and stories with me as part of my doctoral thesis. While their inputs went far beyond the limits of a thesis, I hope this book goes some way in addressing the ongoing technology-augmented injustices and uncertainties faced, resisted, and contested by them. In particular, to E, B, A, and D in New York who shared with me their stories and thoughtfully facilitated my research with their connections and time. To Adriana Norat for your beautiful friendship and for making Brooklyn home for a little while—I wish there were less of a geographical distance between us. For coffee mornings and long wanders with James Duncan. To Mizue Aizeki from the Surveillance Resistance Lab—at the time, with the Immigrant Defense Project—who has since become a close collaborator and co-organizer; to JG for your tireless work fighting and organizing against the technological manifestations of violent border regimes. Thanks especially to Jessica, Nina, Samer, and Dragana for their invaluable time, direction, and critical insights. My admiration and appreciation for the research guidance and opportunities to brainstorm provided by Mark Latonero, Virginia Eubanks, Mark Ghuneim, and Jeffrey Lane. A special thanks to Meg Satterthwaite, whose advice and encouragement was akin to that of an additional adviser. Though I first met them virtually while in New York, I treasure the friendship that emerged on the back of a mutual effort to take techno-humanitarianism to task with A and F. I am also grateful for Lara Nettelfield and Mia White for sponsoring my affiliations with the Institute for the Study of Human Rights at Columbia University and the Zolberg Institute on Migration and Mobility at the New School, respectively.

In Berlin, to B, E, A, and O, who hold a special place in my heart: I am in your debt. For your friendship, knowledge, activism, and resilience. To SB, thanks for having me, thanks for your friendships, and thanks to those who engaged generously with me for the research behind this book.

I've learned a lot, and while some of it was coffee related, a lot more was about urban community resilience and placemaking. A big thanks to Bašak Cali at the Hertie School, in particular, for hosting me as a visiting scholar. To Ben Mason at Betterplace Lab for the resources, connections, collaborations, and comradery in pursuit of our common research interests. To Zara Rahman for being a source of tremendous insights and advice. Thanks as always to Sam Dubberley for the company while in Berlin, the connections, the advice—on matters of research, career, personal, and otherwise—and the solidarity. My unwavering gratitude and dedication to the G100 community in Berlin, who are a fountain of hope in deeply uncertain times.

As a child of a generation of political activists who fled the Iranian regime to start from scratch under conditions unfamiliar, alienating, and—at times—oppressive, I dedicate this work most strongly to my family. My father, a former math teacher who fought against a murderous regime and escaped, worked several odd jobs and retrained in Denmark to provide, for my brother and me, a different future. My mother, a midwife during the Iran–Iraq War, started over in Denmark at a time when women of color were especially subject to significant prejudice and job market discrimination. Despite the severity of their struggles, they imparted, on my brother and me, an urgency in pursuing paths of justice, and intellectual curiosities unfettered by our socioeconomic conditions in the first half of our lives. To my brother—a rock transcendental of geographic bounds—for the humor, encouragement, and sense of wonder with which he has carried himself, and myself when needed, in heavy and light times. To my biological grandmothers in Iran, may they rest in power, who kept me connected with my heritage; and my adopted Danish grandparents, who helped raise my brother and me and grounded us in our confusing cultural hybridity. I fail to imagine a world in which the trajectories of my brother and me would be remotely possible without the strong presence, guidance, and unquestioning and unwavering support of our family. To family members near and far who have inspired me in their courage, thank you with all my being.

To educators who went above and beyond the call of duty to nurture my thinking and for providing opportunities unimaginable. To Ellen Andersen, in particular, who pushed me beyond my comfort zone in defiance of what was expected of Brown kids growing up in the 1990s in Denmark. For supporting and inspiring a truly wild bet that would see me off to Michigan on a six-month scholarship as a fifteen-year-old. My proximity and visits to Detroit left a lasting impression, as the city was crumbling under the financial collapse of 2008, the year of my arrival. There, I witnessed the response of working Detroiters to President Barack Obama's Labor Day rally. I stayed with the Moormans, who not only generously hosted me as if I were their own, but also had a significant hand in informing my politics.

The educators in my life have given me so much, and I've reflected on how I could possibly do them justice. I have come to conclude that there is simply no token of appreciation grand enough. I have leaned heavily on a handful of mentors, intellectual champions, soundings boards, and steady hands both before and throughout the PhD journey during which this book was conceived. In particular, I mean no hyperbole when I say that I am forever indebted to my greatest champion, colleague, friend, and co-conspirator, Ella McPherson, for giving a break to a random undergrad from Queen Mary, when she took me under her wing at the Whistle project, introduced me to the world of digital sociology, and encouraged and supported me throughout my pursuit of an MPhil and subsequently a PhD as a prized adviser. To Graham Denyer Willis, whose pedagogy, empathy, and encouragement compelled me to apply for the PhD program in the first place. His knack for, and apparent superpower, to translate my chaotic mind provided steadiness, guiding me through so many intellectual turns and discoveries for the duration of my research. Graham consistently went all-in for his students, affectionately self-titled the Grahamites. He helped me navigate my often fraught and uncomfortable relationship with academia, and with my work and how I valued it. While postgraduate programs have a way of chipping at one's confidence, I have been honored and humbled to have had a supervisor who built his students up. I only hope that future students of his shall give him less grief—or at the very least, less work in every waking hour of the last months of their research. Furthermore, I cannot thank him enough for acting as de facto academic matchmaker, pairing me with my research sparring partner of nearly five years, Giulia Torino. In turn, all my admiration, gratitude, and joy for Giulia for diving headfirst into academic insanity, writing anxiety, submission planning, existential dread, pandemic panics, job applications and motivation with me, on a far too regular basis.

I am equally indebted to colleagues turned co-authors and friends from the Whistle and the Centre of Governance and Human Rights, which became intellectual homes of sorts, providing tremendous support, guidance, and allyship during my time in Cambridge. To Sharath Srinivasan, Isabel Guenette Thornton, and Bekah Larsen in particular. Importantly, I want to acknowledge that this work would not have been possible without the love, fierce friendships, regular sanity check-ins, companionship, peer review, and world-building of so many beautiful people. Thanks to our transatlantic "fefe club," Emma Murphy and Elise Williams. To Luke Naylor-Perrott and Natalia Hussein, Damini Satija, Jay Richardson, Theo Weiss, Ali Neilson, Lucie Vovk, Inès Aït Mokhtar, Rebecca Vaa, Alia Al Ghussain, Sophie Dyer, Max Kashevsky and Naomi Marshall, Niyousha Bastani, Corinne Cath-Speth, Tellef Raabe and Anna Eitrem (and little Eva, and soon Lillebror), Aurelie Skrobik, Danielle Cameron, Sahil Shah, Ailsa McKeon, Surer Mohamed, Cansu Karabiyik, Lili Pierret, Marlena Wisniak, Sam Dubberley, Waed Abbas, Victoria Tse, Rohan Clarke, and Alex Grigor. To you all, thank you. We have shouldered everything between tears, laughter, despair, and joy together. Neither would it have been possible without my broader Cambridge family, and home away from home, in particular Helen Jennings, Jana Weber, Selina Zhuang, Ben Jackson, Lorena Gazzotti, Maria Hengeveld, Jane Darby Menton, Rafaelle

Danna, Joe Shaughnessy, Lena Moore, Emilio Garciadiego, Barry Colfer, Sophia Borgeest, Noura Wahby, Jonathan Woolley, Georgiana Epure, and Nikta Daijavad.

To former and current partners and organizers-in-crime at No Tech for Tyrants and at Amnesty Tech. To Mallika Balakrishnan, Quito Tsui, Jacob Zionts, Sophie Dyer, Rasha Abdul Rahim, Michael Kleinman, Charlotte Phillips, Likhita Banerji, Verity Coyle, Abigail Robinson, Maen Hammad, Saleh Hijazi, Justin Mazzola, Tom Mackey, Marija Ristic, and Milena Marin. Thank you, you're all an inspiration.

Over the years, I have also benefited tremendously from mentors and intellectual sparring partners without whom this research would not have come together in the way that it did. In particular, Alix Dunn, Kira Allmann, Ilia Siatitsa, Alexandra Winkels, Raymond Apthorpe, Corinne Cath-Speth, Antonella Napolitano, Bushra Ebadi, Gabi Ivens, Jennifer Cobbe, Alexa Koenig, Monica Greco, Elizabeth Eagen, and Matthew Burnett.

While at Cambridge, the two most consistent channels of creative work and support have been, firstly, my "Declarations" family over its many generations, including Scott, Talia, Eva, Max, Niyousha, Surer, Helen, Jing, Genevieve, Arindrajit, Daniel, Misbah, Jonas, and Michael. Secondly, our transnational DVC family, especially Ray, Dylan, Adebayo, Haley, and Nickie. Toward the end of research, I was in the good graces of Sarah Villeneuve, Lucie Vovk, Asher Goldstein, Danielle Cameron, Jasper Tautorus, Victoria Tse, Niyousha Bastani, Sam Dubberley, Rebecca Vaa, and Aurelie Skrobik, who proofread my classically delayed chapters, as they arrived in their inboxes frenetically and at ungodly hours. I am indebted to you for all your troubles and am eternally grateful for your time and effort. Thank you.

This book owes its existence to the kind encouragement of Ali Meghji and Angèle Christin, whose generous feedback and direction helped the work on its way from its initial form as a PhD thesis. Ali, Ella, and Graham all subsequently provided invaluable insight and advice, as I navigated the world of academic book publishing, while the book benefited from the input of two wonderfully constructive reviewers, including the inspiring Eleanor Drage. A huge thanks to Naomi Schneider and the incredible team at UC Press, including Aline Dolinh and Emily Park, whose guidance and championship I remain ever grateful for.

To the ever-wonderful Amy Costelloe, you have shown up and held a light in ways the acknowledgments section of a book is ill-suited to describe—thank you.

Finally, this book was completed in the spirit of Jo Cox, MP—the namesake of the PhD studentship I was endowed with at the time of this research—who was killed on June 16, 2016, by a far-right activist for her advocacy on behalf of refugees in Europe. The financial support of the Jo Cox PhD Studentship at Pembroke College, University of Cambridge, and the Cambridge Commonwealth, European and International Trust made this research possible. My fondness of Pembroke reflects the wonderful concoction of academics, staff, and students, my engagements with whom have impacted me greatly. In particular, I am grateful to James Gardom, Loraine Gelsthorpe, Nami Morris, Emily Hinks, Sanne Cottaar, as well as the encouragement of Chris Smith.

Acronyms

BAMF	Bundesamt für Migration und Flüchtlinge
DOS	US Department of State
HPD	NYC Department of Housing Preservation and Development
ICoTs	Information control technologies
ICTs	Information communication technologies
MOIA	Mayor's Office of Immigrant Affairs
OMNY	One Metro New York
R&P	Reception and placement
UNHCR	UN High Commissioner for Refugees
USCRI	US Committee for Refugees and Immigrants
VolAg	Voluntary agency

Glossary

DATAFICATION An idealized attempt to render raw and neutral representation of human life and social problems in purely numerical terms.[1]

DIGITAL INFRASTRUCTURE Digital structures, facilities, networks, and tools that allege to enable, for example, access to services, communication, and other essential means to move in society.[2]

DIGITAL PERIPHERY The digital categorization and containment of historically racialized and/or colonized communities for technological exploitation in service of capital and prestige.

DIGITAL URBAN INFRASTRUCTURE Digital infrastructures for accessing information, networks, services, and socioeconomic life in cities, with consequences for how bodies and movement are ordered.

INFRASTRUCTURE The basic structures, facilities, and networks needed to operate a society or organization, for example, the services, roads, and buildings in cities or, alternatively, the cloud servers, data centers, and algorithms needed to run artificial intelligence tools.

RACE Social construction intended to categorize people by phenotypical difference, typically with associated tropes rooted in hierarchies of power.[3]

RACIAL CAPITALISM Contemporary economic world system, originating in the seventeenth century, in which racial hierarchies are exploited in service of capital.[4]

RACIALIZATION The process of imparting racial characteristics and meaning to an individual, group, and/or practice.[5]

RACISM Attitudes, approaches, behaviors, policies, and regimes at large, which reify hierarchies of social relations emanating from proximity to whiteness and, by extension, implicit or explicit superiority.[6]

SMART CITY A city in which data-intensive technologies, including artificial intelligence, biometric recognition, and other digital tools, are seen as primary solutions to major urban social, economic, political, and environmental problems.[7]

SOUSVEILLANCE Also known as surveillance from below, sousveillance is typically associated with surveillance activities directed at those in power, by those conventionally subject to state surveillance, for example.[8]

SURVEILLANCE Observing mechanisms of social control that target individuals and/or groups, encompassing their living areas, behaviors, and daily activities.[9]

INTRODUCTION

They pretend they're just doing market saturation but then they fucking depend on government contracts! Like we're back in the nineties. . . . Your neoliberal bullshit doesn't confuse me.

AN ORGANIZER WITH MIJENTE,[1] a political home for Latinx and Chicanx people who seek racial, economic, gender, and climate justice, describes how the organization discovered that the horrifying regime of deportation raids and family separations advanced by US Immigration Customs Enforcement (ICE) had been powered by Silicon Valley companies. At the time of our conversation, Mijente had just launched its #NoTechForICE campaign,[2] on the back of a tell-all report documenting how household companies like Microsoft and Amazon were the technological engines powering ICE.[3] More insidiously, companies like Palantir Technologies, a Peter Thiel and Alex Karp venture made infamous by their initial CIA seed investment,[4] have been providing software that directly aided in ICE's targeting and detention of undocumented communities.

"Yes! No one got arrested!" they exclaim at an incoming text with indignation. #NoTechForICE organizers had erected a chain-linked fence around the Palantir offices that day—a reference to the egregiously barren detention centers, and the deadly border-wall, to which the company was in effect condemning marginalized migrants. For Mijente—and many other advocacy groups concerned with immigrants' rights—companies like Microsoft, Amazon, and Palantir are the future of incarceration. Their rise to prominence and entanglements in matters of security, incarceration, and surveillance marks a monumental shift in the course of migration governance. While unprivileged movement remains a lethal endeavor for a great many communities in the world, others profiteer from their precarious existence and the possibility of their movement.

As the fortification of Europe's borders and its hostile immigration terrain—often referred to as Fortress Europe—took shape, so the biometric and digital surveillance industries also grew. And when ICE aggressively reinforced its program of raids and egregious forms of detention and family separation, it was powered by Silicon Valley corporations, which provided the technology for tracking, case management, and algorithmic decision-making. Even in cities—historically sanctuaries for refugees, asylees, and undocumented immigrants alike—where communities on the move have lived in anonymity and proximity to familial and diaspora networks,[5] the possibility for escape is increasingly diminishing. As cities of refuge rely increasingly on tech companies to develop digital urban infrastructures for accessing information, services, and socioeconomic life at large, they are also inviting the border closer to cities and even closer to migrant communities, their networks, movement, and bodies. This marks a convergence of Silicon Valley logics, austere and xenophobic migration management practices, and racial capitalism. As actors involved in the regulation of movement, and therefore in migrant lives, technology companies are closing in on the final frontiers of fugitivity to datafy and commodify those escaping their circumstances in search of something better.

ALGORITHMS AS BORDERS

Imagine narrowly escaping death, battling the tide, hostile coastguards, and drones across the sea. Even before you arrive, somewhere, on some anonymous server, you are red-flagged by an automated decision-making system. At the border checkpoint, a massive database of preregistered fingerprints confirms the identities of incoming migrants, yet border guards demand that every face be scanned, ostensibly for more efficient and accurate processing. Every processed face is piled together with *other* undesirables in line for human processing. A predictive algorithm processes your mobile phone—acquired while on the run—as you stand in line at the port of entry and warns the border guard not to trust you, that you're from a country you only transited through for a few hours. You're flagged for emotion recognition and extensive interrogation. Imagine arriving to your city of destination months later, only to be red-flagged by yet another algorithm, which notifies the authorities that you are scheduled for deportation. You avoid the safety of your usual route to work because the path is littered with facial recognition and sensors that passively collect data from your phone. You delete the apps you were recommended to get by in the city, having learned that they snitch on you—that one neighbor down the road had been sent to a detention center in another state because of them, another denied health care. These are just some realities of communities living in the digital periphery.

The lines between borders and border subjects, such as refugees and migrants, have increasingly blurred. A bizarre and futuristic combination of biometric identification,[6] such as fingerprinting, iris scanning, voice and facial recognition, and even DNA collection, has become part and parcel of the tools used against communities on the move.[7] For instance, following return, Afghan refugees have faced compulsory iris registration programs, administered by the United Nations High Commissioner for Refugees in order to receive assistance. In

Jordan, the World Food Program deployed the EyePay program, supplied by the company IrisGuard, for the Za'atari refugee camps, ostensibly in an effort to counter fraudulent behavior among residents. Humanitarian workers who had previously championed the program had meanwhile sounded the alarm; not only was there no data to support the idea that fraud was a significant problem in the camp, but incoming refugees were disincentivized from registering in the camp in the first place and forced to forego life essential goods and services to avoid biometric conscription. On the US southern border, technological fortification since the 1970s has meant the installation of intrusion detection systems,[8] precursors to the now dozens of surveillance towers it is equipped with, set up by Israeli surveillance company Elbit Systems for use by ICE "to improve border security and force protection" with the help of "artificial intelligence and automation."[9]

While global movement and information technologies are changing practices of bordering, such matters are now also substantially urban. Today, refuge is datafied and transcendent of the borders drawn around conventional colonial geographies. Increasingly, the same processes by which displaced populations are assessed, afforded access and information, and surveilled between borders now exist within borders, states, and cities as well.[10] These processes mask, for marginalized mobile populations in particular, the mutually constitutive relationship between race, borders, and migration by way of technologies that datafy, automate, and further scale colonial relations.[11]

In recent years, rapidly digitalizing cities, some with explicit "smart city" agendas, have developed digital strategies directly or indirectly aimed at the integration of migrant populations. Cities and local authorities have responded to the presence of immigrant communities by deploying digital technology interventions, purportedly to mitigate access to crucial information and services. New York City is one such example, where the Mayor's Office of Immigrant Affairs (MOIA) has had a direct approach to providing a digital strategy aimed at communities on the move (translation services, immigration services, connectivity, identification). In an environment prone to deportation raids, infrastructural technologies such as sophisticated public Wi-Fi, smart ID cards, and digitalized city services have led to deep anxieties among refugees, asylees, and undocumented immigrant communities related to information-sharing, surveillance, and retrievals. Immigration advocates doubt the compatibility of New York's sanctuary city status with its aspirations as a so-called smart city, due to uncertainties surrounding the collection, use, and sharing of resident data by the city's digital services and surveillance infrastructure. Indeed, as will be explored in chapter 4, New York's extensive surveillance capabilities compound antisanctuary logics in two ways: First, through generating information panics; that is, forms of informational precarity that disincentivize affected populations from engaging with city services. And second, through the instrumentalization of the smart urban milieu for immigration enforcement. While information panics reinforce the antisanctuary through further coercing communities on the move into urban concealment, the instrumentalization of city infrastructures expands the field of vision for adversarial immigration authorities such as ICE, US Customs and Border Protection, and the New York Police Department.

Berlin is another such city—here, a combination of civil society and smaller private-sector initiatives purport to integrate refugees. The more than twelve million refugees who have fled the Syrian crisis have provided especially ample opportunity for technologists seeking to test and develop tools for access to work, housing, and credible identities.[12] Civil society and private-sector initiatives target refugees for these services, and scholarship has, until recently, tended to treat positively the affordance of the smart city for refugee integration.[13] However, these developments reflect existing racialized integration narratives and the construction of a precarious urban–entrepreneurial refugee "periphery." As we'll uncover in chapter 5, Berlin's digital refugee response is part and parcel of a form of digital ghettoization, in which digital refugeeness is used as a symbolic form of capital to attract funding while enabling tokenistic claims of political solidarity.

For targeted communities on the move, the digital city is inextricable from the emergence of a digital urban border. As populations who do not possess the same level of protection as citizens,[14] displaced communities are at risk of being used as experimental sites in pursuit of racial capital.[15] As we begin to understand these technologies to be constructive of borders, it becomes clear that these tools—be they apps, biometric technologies, algorithmic decision-making, or artificial intelligence more broadly—do not end with the product itself. The fetishization of digital tools for development and deployment in the name of historically marginalized communities enables the automation of exclusion. More recently, scholars have begun to develop a crucial vocabulary around these forms of fetishization: Meredith Broussard has referred to it as "techno-chauvinism" (the insistence that technology is better than the human)[16] and Evgeny Morozov as "techno-solutionism" (the idea that there is a technological solution to the most complex societal problems).[17] Far from a new phenomenon, the use of technical language and numbers to advance an illusion of control is actually a long-standing feature of colonial conduct: from the nineteenth-century birth of the British census in India as an attempt at exerting atomized social control over its colonized subjects, early eugenicist classifications of intelligence (IQ), the IBM Hollerith punch-card system used to systematize and manage concentration camps, and symbols and letters drawn on the skin of immigrants arriving at Ellis Island right through to the ethnic and racial classifications in personal identification cards in South Africa and Rwanda.[18] These—often lucrative—modes of subjugation were employed under the guise of technical or scientific fixes to various problems associated with largely racialized communities. Boomeranging back to our contemporary predicament, technology deployments in the context of the so-called global refugee crisis repackage similar colonial imperatives for today's liberal imaginaries.[19] That is why I insist on taking a point of departure in racial capitalism and practices of bordering as the organizing structure behind technology production in the digital age; borders, and now cities, have become experimental sites where populations are digitally enclosed, and where life is subject to exploitation by tech actors.

GLOBAL RACIAL CAPITALISM IN SEARCH OF A SCAPEGOAT

In the United States, between 2014 and 2019 alone, nearly two thousand individuals died along the US–Mexico border,[20] while 210 individuals have lost their lives in the custody of ICE.[21] Global flows of human movement are animated by these numbers, which show that not all movement flows equally. Communities fleeing dire circumstances of various kinds are required to pay with their movement; at the same time, their movement is relied on by technology companies for profit, what Todd Miller has called the border industrial complex.[22]

In November 2020 the UN special rapporteur on contemporary forms of racism, racial discrimination, xenophobia, and related intolerance released her report on race, borders, and digital technologies, urgently noting that

> the resurgence of ethnonationalist populism globally has had serious xenophobic and racially discriminatory consequences for refugees, migrants and stateless persons. . . . Digital technologies are being deployed to advance the xenophobic and racially discriminatory ideologies that have become so prevalent, in part due to widespread perceptions of refugees and migrants as per se threats to national security.[23]

Achiume's report underscores the conundrum we face: a slew of racially coded and discriminatory technologies heralded as a solution to the crisis of moving communities. Even the so-called crisis is not one of migratory makings, but one of global racial capitalism in search of a scapegoat. The very consensus between technology actors and cities to share in the management of migrant life and survival in increasingly smart cities must be examined against the backdrop of what neoliberalism looks like in the twenty-first century. Under this paradigm, xenophobia begets technology deployments under the guise of greater control for state authorities, which in turn begets profits for tech companies; meanwhile, resources remain above the heads and out of sight for both those conscripted into cheering on, and those being accosted by, the hostile immigration environment.

For example, I arrived in New York City in a moment of particular turmoil for immigrant communities at large; US Secretary of Homeland Security Kirstjen Nielsen had signed off on "option 3" in April 2018—a deterrent strategy set to curb migration by putting in motion the indiscriminate prosecution of "every adult who crossed the border illegally, including those who came with children."[24] DHS ramped up its now-notorious deterrent policy of family separations, and it also emerged that families were unable to reunite due to technical shortcomings that prevented parents from being digitally "tagged" with their separated children. While the Trump administration was by no means the first to boast an aggressive deportation machine, it did become the first in US history to have birthed the concept of "deleted families."[25] Trump's border didn't simply end with his infamous wall or in immigration courts—the border was now capable of digitally mediated migrant erasure altogether. The border, in other words, "transgresses" families and fortifies the artificially imposed divides between them.[26] The digital augmentation of the US border, however, did not emerge overnight, nor was it limited to the purview of border enforcement agencies such as ICE and US Customs and Border Protection.

With 343 deportation raids between January and October 2018 alone (at least one raid per day on average),[27] the irony of New York's sanctuary city status did not escape the many caseworkers and immigrants' rights activists I encountered in the city. Moreover, the term itself hardly featured in how asylum seekers, asylees, and other immigrant populations conceptualized their presence in New York. Although the sanctuary city exists on two levels—that is, as a movement driven largely by civil society and religious organizations,[28] and as a policy adopted by MOIA[29]—it also remains the administrative, material, and spatial embodiment of violent gatekeeping practices, contravening any potential right to the sanctuary city.[30]

In this book we will grapple with two dynamics that are increasingly normalizing violent technology deployment against moving communities at scale. First, how such technological interventions are seen as a way of bypassing the politics of migration and integration (meanwhile deferring power to but a few large tech corporations). And second, how inequities and discrimination resulting from these interventions have been defined as technical questions that could be surmounted if only we had more diversity among technologists and engineers, if only we had more varied data, or if only more institutions would use our systems. These retorts are different sides of the same techno-solutionist coin, which posits that the answer to stopping the violent consequences of racial capitalism is to be found among the perpetrators themselves. Meanwhile, partnerships spearheaded by technology giants in service of humanitarian and international organizations and backed by massive philanthropic institutions have experimented with interventions ranging from blockchain-based identity systems to algorithmic resettlement schemes. These very actors have also turned around and provided the same technological infrastructures they are using to purportedly enable mobility in one place to facilitate the detention, deportation, and separation of families in another.[31] In the chapters that follow, we will expand our conventional understanding of the border to encompass the urban, where new forms of digitally mediated enclosure are utilized for migration control, to the effect of justifying the experimentation and development of technology products on the backs of vulnerable displaced communities.

WHAT IS THE DIGITAL PERIPHERY?

Nick Couldry and Ulises Mejias have referred to the process by which "many specific aspects of human life [including] the grid of judgment and direction that we call 'governance'" are appropriated as "data colonialism," in particular because it enables an externally driven "appropriation of data on terms that are partly or wholly beyond the control of the person to whom the data relates."[32] In 2018 Safiya Noble published her pioneering work *Algorithms of Oppression,* which traced the racial logics underpinning search engine algorithms,[33] while Cathy O'Neil and Virginia Eubanks respectively, both examined how preexisting socioeconomic inequities are augmented and amplified through algorithms.[34] More recently, Ruha Benjamin has referred to the machinery behind how carcerality is replicated digitally as the "New Jim Code."[35]

The last decade has seen an explosion in tech-critical vocabulary to help us understand and contest these developments,[36] in particular to make sense of how sophisticated technologies are used in regimes of control. Emergent works on the centrality of cities to technology development and deployments have also begun to appear,[37] along with some exploration of navigating resettlement in cities through digital technologies.[38] In 2019 Shoshana Zuboff gained notoriety for her framework of "surveillance capitalism" through her book by the same title.[39] Yet surveillance capitalism, as a framing, falls short of contending with not only capitalism itself but also its racial underpinnings. Critical scholarship from the last decade has also given much credence to the exploration of algorithmic racism and bias as technical errors to be fixed, in studies ranging from bias in facial recognition systems[40] and recommendation algorithms[41] to welfare assessment algorithms and beyond.

However, insisting that these rampant practices of the datafication of the everyday—and subsequent sales of that data—are merely capitalism "gone rogue" ignores how racialized surveillance and control have been core to the very formation and function of racial capitalism. Similarly, an emphasis on technical biases in such technologies runs the risk of obfuscating effects at the core of racial capitalism, attempting to debias systems that are fundamentally operating as intended. Work by scholars including Meredith Broussard has gone some way in explaining that discrimination along ableist, gender, and racial lines is the result of not design glitches but "bugs" innate to human beings, which therefore naturally and ultimately feature in their technological output.[42] To understand how cities become sites of racialized migration control, *Migrants in the Digital Periphery* explores how the current trajectory of the tech industry is by no means an exception or "perversion" of capitalism,[43] but follows instead a rich historical genealogy tied to the control and exploitation of the mobility of racialized communities: control and exploitation as a *feature*, not a bug.

Similarly, works that examine increasingly sophisticated police usage of technology, in particular artificial intelligence, have tried to make sense of how such systems—what Frank Pasquale has called the "black box"[44]—operate in technical terms.[45] In *Predict and Surveil*, for instance, Sarah Brayne studies police uses of big data and nascent surveillance technologies.[46] She interrogates police departments' negotiations with problematic tech actors, who woo them with promises of progress and cost-efficiency. While Brayne's work crucially takes a point of departure in ethnographic encounters with police officers and those who procure, learn about, operate, and negotiate big data policing, *Migrants in the Digital Periphery* looks at how similar and adjacent technologies are experienced and described by the communities subjected to them. In this book, I have also deliberately veered away from making any serious claims about the inner technical workings of the so-called black box. Instead, I have sought to locate how these technologies operate on marginalized subjects in particular, experientially, through the active witnessing of how these systems have shaped the everyday experiences of life in digital cities of refuge.

In works taking borders to task for their violence in a time of technological acceleration, the focus has tended to be on how the physical border is propped up with artificial intelligence, drones, predictive analytics, and other remote-sensing capabilities,[47] or how refugee camps have been turned into laboratories for technological experiments.[48] This work matters greatly; without these crucial contributions, we would be short of documentation for a whole category of grave injustices emanating from Silicon Valley firms and those fashioning themselves like them. However, less attention has been paid in these works to the ways camp-like conditions, and indeed borderization, continues to endure beyond the camp and beyond the border through technological interventions.

Zooming in on today's modes of control and extraction in service of capital as somehow historically unique misses the bigger picture and distracts from the racialized logics driving the tech industry today. By paying attention to contemporary tech deployments from the vignette of racial capitalism, we realize instead that the treatment of *algorithms* being racist reduces violent racial outcomes to excusable programming mishaps. This, in turn, normalizes the undergirding system—the actors, institutions, political and economic logics—that animate these products and, by extension, their violence. Conversely, and as the decolonial computing scholar Mustafa Ali, argues, racism itself is programmatic or "algorithmic,"[49] and this undergirding logic has historically given (and continues to give) impetus to a series of logics that undergird global processes today.

With this book, we will explore how technological processes—especially those deployed under the auspices of alleviating conditions of marginality—are never about the technology itself, but about how alterity and race are constituted, innovated, and weaponized in service of racial capitalism. This plays out in three major ways: practices of digital bordering follow a long tradition of racially underpinned enclosure, containment, and value extraction; practices of digital bordering go beyond material borders and seep into the realm of the everyday; and practices of digital bordering are *not* unintended externalities or ethically rectifiable processes but symptoms of colonial continuity.

This book emerges on the back of doctoral research I conducted across New York and Berlin, initially between 2018 and 2020, with follow-up research between 2020 and 2023. By working in active solidarity with my participants in uncovering and getting "beneath the surface of oppressive structural relationships,"[50] paying particular attention to the sharp of technology as ideology and the system of racial capitalism, I take a leaf out of the critical orientation of the ethnographic tradition. This, in turn, enabled me to study mundane digital infrastructures that have relevance to aspects of urban life among displaced communities in New York and Berlin to develop a better understanding of how social exclusion is experienced and contested. This blended design approach allowed me to engage in predominantly short- to medium-term "bursts" of fieldwork, during which both longer-term participant observation (four to five months in each site, documenting everyday interactions, conversations, and observations using field notes) and short-term engagements, such as one-off interviews, workshops, and other events, informed my research.

I spent the months of August 2018 through January 2019 in New York, where I was embedded within a resettlement and economic development organization in Flatbush, Brooklyn, with a particular focus on immigrant communities. At the organization—which I refer to as Reset to preserve its anonymity—I assisted caseworkers in providing support services, including resettlement and workforce development support, while researching available housing options amid a deepening crisis of space in the city. I was primarily tasked with setting up a centralized resource for alternative—including community-based and informal—forms of housing, to aid Reset in making the case for continuing its provision of resettlement and placement and advancing its goal of achieving "self-sufficiency within first six months" for refugees. Reset was interested in potential technology solutions for this. Through the prism of this objective, I came to encounter the depths of hostile immigration practices in New York and how they were digitally enforced (what I have subsequently referred to as the digital antisanctuary). Through my quotidian engagements with Reset and thirty semistructured interviews across six additional immigrants' rights organizations, technologists, and the MOIA, I repeatedly came up against anxieties, among refugees, asylees, and undocumented immigrant communities, related to the sharing, surveillance, and retrieval of information.

I was based in Berlin between February and June 2019, where I divided my time across two organizations. I became a volunteer at a collective built in the image of South African sharing houses, which were originally intended to house marginalized Black populations and support them in their activism; I have referred to this as Sharehouse Berlin throughout the book, to preserve anonymity. Sharehouse worked on behalf of refugees and newcomers. Individuals and young families from Sudan, Somalia, Iraq, Afghanistan, Syria, and Iran live in the West Berlin building's upper floors, while the ground floor is used as a café and event space to make up the cost of living. From here, I was introduced to many newcomers and newcomer-led initiatives who ended up providing crucial insight into the techno-politics of refugee life in Berlin. Techfugees is a multichapter organization that purports "to empower the displaced with technology."[51] I joined the Berlin chapter of Techfugees as a researcher, during which time I observed and documented the sociotechnical practices and logics of the organization. I also relied on the partnership of Betterplace Lab to develop a comprehensive overview of digital technology interventions for refugees, and the newcomer-led research initiative G100, to map said interventions against newcomer priorities. I conducted twenty-seven semistructured interviews across ten "refugee tech" initiatives and research and refugee rights organizations.

Taking a point of departure in the rapidly digitalizing sanctuary cities of New York and Berlin, this book traverses the stories of self-identifying immigrant communities and their encounters with digital urban technologies initiatives developed in their names. In these cities, I try to make sense of how practices key to the development of racial capitalism work together on historically marginalized subjects to form digitally mediated enclosures for value extraction. To understand this, I also grapple with some of the earliest historical drivers of racial capitalism. For example, how from its formation into the digital age racial capitalism has relied on two key modes of subjugation: categorization and containment. Categorizing colonial subjects creates an illusion of control and ownership that can be exerted on racialized communities. Through containing (or enclosing) categorized communities in space to work land as indentured servants or extorted laborers, value can be extracted and turned into profit.

The datafication of displacement—and indeed of racialized communities in marginalized contexts in general—over the last two decades has replicated and augmented these modes. Individuals and communities are now categorized under the flattened digitalized categories of refugee, Black, unintegrated, irregular, and developing world. And the insistence on their immobility (being confined to a camp, a detention site, or unable to move outside particular areas out of fear of persecution and deportation) leads to their exploitation for the deployment of experimental technology interventions. This is the digital periphery, where tech companies like Meta (formerly Facebook), Microsoft, and Google use emancipatory language to situate themselves as political stakeholders working toward justice-oriented goals. They partner with governmental and nongovernmental organizations alike to gain new ground for technology deployments.

In a moment when sanctuary cities are once more becoming relevant spaces for organizing, against the backdrop of rising anti-immigrant sentiment, the pages that follow unveil how digital urban infrastructures interact with practices of racialized bordering. This book highlights how moving communities—refugee, migrants, asylees, and otherized people fleeing racial regimes—as well as their bodies and identities, become contested spaces in the battle for racial capital. They have been transformed into frontiers in which technology actors are chiefly concerned with reconstituting conceptions of race—often in service of governments and institutions—and in so doing, moving the needle of permissibility in acts of experimentation, exploitation, and violence. This book, in other words, sketches how urban processes are intertwined with constellations of global human movement and the imperative to govern and profit from it.

CHAPTER OVERVIEW

Chapter 1, "Racism Is a Feature (Not a Bug)," unpacks the historical underbelly of computational power and tech development. I trace how the predominant modes of racialized subjugation from the earliest days of the formation of racial capitalism, namely categorization and containment, can be traced historically, and how they undergird tech deployments today. Here, we look to understand how race, mobility control, surveillance, and extraction are design features of capitalism, not bugs.

Chapter 2, "The Making of the Digital Periphery," delves deeper into the concept of the digital periphery as a framework for interpreting how racial capitalism works in the digital age. The digital periphery is mobilized in three major ways by those who benefit from it: namely through techno-development, techno-space, and techno-government, which can be thought of as three distinct forms of bordering. It operates techno-developmentally—that is, through the instrumentalization of aged colonial tropes by tech actors, inherited from a long tradition of Western representations of "the needy," particularly in the humanitarian and development sectors. Techno-spatially, I show how city-dwelling migrant communities are conditioned for persistent precarity through the threat of surveillance, which forcibly contains them in physical and virtual space for the continued deployment of tech on them. Finally, the digital periphery matters techno-governmentally; governments and institutions rely on technology actors and their digital urban interventions, symbiotically, to identify, exploit, and control immigrant communities.

Chapter 3, "Xenophobic Roots, Tolerant Facades," takes the ideal of the city of immigrants or sanctuary city to task, tracing some of the most consequential historical anti-immigration laws and policies in New York and Berlin. In New York, processes of assessing arrivals into the port have become diffuse throughout much of the twentieth century, steadily moving away from the port and disaggregating into everyday urban life—a condition that made the city ripe for a digital form of antisanctuary. In Berlin, and Germany broadly, the history of the immigration policy landscape complicates the much-popularized policy of *Willkommenskultur* (culture of welcome) that emerged at the height of the Syrian crisis in 2015. This is a context that has largely refused stable routes to permanent residence for non-Western immigrants in particular until recently, when pressure to recruit migrant labor for the tech sector started piling on the German government.

Chapter 4, "The Digital Antisanctuary of New York," takes a point of departure in New York, a city that has become a site of mundane border enforcement beyond the border, owing in part to its increasingly digitalizing infrastructure. Here, we meet migrants and privacy rights activists, public defenders, social workers, and technologists who tell the multifaceted story of how urban migrant environments have become commodified and datafied. The deference of cities like New York to technological solutions in realms crucial to everyday life, such as access to information, identification, and housing, permits technology giants to play a subtle yet increasingly active role in the control of undesirable migrant bodies. In exchange, cities such as New York can continue posturing as "sanctuaries," while facilitating the rapid and lucrative entrenchment of Silicon Valley in the fabric of urban governance. Here, I argue that the city is a legitimizing ground for the Valley and a techno-purgatorial containment zone for those fleeing persecution and hunger.

Chapter 5, "Digital Refugeeness in Berlin," documents how the increasing availability and usage of apps that target refugees for access to information, services, work, and housing potentially transforms how refugees access life in the city. With little to no oversight and accountability, this neoliberal approach to urban refuge in Berlin perpetuates the deep-seated myth that refugees and vulnerable migrant populations are made of fundamentally different matter—that their needs, literacy, and even desires around flexibility and stability are distinct and by nature more precarious. This, in turn, gives rise to digital refugeeness. Digital refugeeness does not depend on demonstrating relevance or benefit for the communities it encompasses in order to exist; in fact, it exists almost purely for the enjoyment of experimentalism and the solicitation of technical and financial capital between urban entrepreneurs, larger tech companies, and governmental as well as nongovernmental institutions. Contrary to New York City, the technology interventions in Berlin do not necessarily sustain surveillance structures related to immigration enforcement in the city; nevertheless, they commodify and datafy their identities' subjectivities. The encounters in these chapters shed light on the workings of the digital periphery, and how seemingly disparate and decentralized forms of digital socioeconomic interventions convert the refugee's predicament into a laboratory.

Chapter 6, "Disciplining Mobilities in the Digital Periphery," outlines how technology initiatives in both New York and Berlin have given rise to information panics and system aversion, which further compound precarity and insecurity. Notably, initiatives such as LinkNYC and IDNYC contain individuals either through detention and incarceration or keeping them fixed in urban space through information panics, thereby categorizing them for subsequent deportation. In Berlin, newcomer identities have been exploited to keep start-up capital flowing between funders, the state, and technology initiatives, at the expense of reinforcing a racialized assimilationist framing around refugees. This chapter takes stock of how communities resist, evade, and refuse these interventions and their underlying xenophobic politics. Reflecting on the involvement of tech companies in US immigration enforcement, my informant from Mijente succinctly describes Silicon Valley's market penetration by saying, "It's their innovation curve: start with war, then refugees, then the mass."

In the conclusion I look at how campaigns such as Close the Camps, #NoTechForICE, RethinkLinkNYC, and the Immigrant Defense Project have taken direct action at the offices of technology giants such as Google, Amazon, and Palantir. These campaigns build on a rejection of the "smart" altogether. They present a break with approaches that both seek to improve the experience of marginalized populations with technology deployments and assume the inevitability of tech. Their refusal, whether in the form of urban concealment, the use of alternative technology configurations, or physical tech sabotage, presents a critical neo-Luddite epistemology of migrant survival in digital cities of refuge.

Finally, I suggest that in their nonuse and refusal, communities confer meaning and reveal partial truths about digital urban infrastructures, the interests that operate them, and the sources of authority that communities reject and dismiss in so doing. These communities present a possibility for change and a vision that starkly contrasts and negates the techno-solutionism and -chauvinism that has been characteristic of Silicon Valley for much of the last two decades. Displaced communities, immigrants' rights advocates, and privacy activists in both New York and Berlin have sought to challenge technology deployments targeted against migrants in particular, on the grounds of their potentially devastating consequences for the communities they purportedly intend to serve.

PART ONE

RACE, BORDER, AND CAPITAL ENTANGLEMENTS

1

RACISM IS A FEATURE (NOT A BUG)

I DREAD THIS PART EVERY TIME. The minutes spent wondering whether you should smile or keep the awkward blank look that even you, within yourself, find utterly suspicious. The cyclical thoughts that hope to compel a flippant response to your practiced greeting, delivered at the exact time of day to demonstrate that you are literate and therefore more of a citizen than your skin might otherwise reveal, and in as formal a way as possible, lest they catch you out for being too "ghetto." The minutes that pass while you practice your defense to a yet-unknown offense, while the border guard inspects your passport as if to emphasize his dismay with how atrociously different it looks from the passport he inspected just a little while earlier (it does not). Then the "hold up": when he holds up your passport next to your face, puts it back down to reexamine not only if the passport photo fits your face but if the passport is real (despite having swiped it through the system successfully). With the "wrong" last name, you could end up in a backroom, stuck in border limbo simply because you . . . exist. While physical border checks are eerily othering experiences under which every inch of our personhood is questioned, they bring materiality to the border; you see it and its racialized characteristics in front of you. However, since the early days of the war on terror, borders have been given new characteristics,[1] operating in the digital realm as images, probabilities, flags, and algorithms that follow you. Tech companies are now lucrative border-builders, fusing our bodies to borders through technologies like facial recognition, lie detection, predictive policing, and so much more. And borders, in turn, have become elastic, tightened to different levels of tension, ready to snap you into limbo at any given moment. Ours is a world obsessed with borders. Lives are made and unmade by their situatedness to borders; news and politics is curated on the back of border crises, which in turn drive much of twenty-first-century ideologies of security and capital.

Contemporary processes of mobility management, including their technological manifestations and appendages, are mundane forms of colonial continuity. And while the technology-augmented violence of these practices is not always clear (against the backdrop of daily news report of mass displacement, drownings, killings, and other forms of death), the entrapment of communities on the move in exceptional categories—for the justification of tech development and deployment—is in itself a death sentence of sorts. We need a conceptual framework that allows us to glean these dynamics. To fully grasp how smart cities impart this particular flavor of violence, we must prime ourselves to perceiving algorithms—and by extension the digital urban infrastructures they constitute—as borders. In this chapter, I unpack the historical underbelly of computational power and tech developments, and trace how racialized forms of categorization and containment undergirding tech deployments today have a rich history, derived from the earliest days of the formation of racial capitalism. This is an important tool in understanding how race, mobility control, surveillance, and extraction are design features of capitalism, not bugs.

MUSLIMS, REFUGEES, AND TERRORISTS? THERE'S AN APP FOR THAT

In Europe, billions of euros have been invested in bolstering the capacity of the border agency Frontex to surveil, register, and track moving communities. The agency is directly complicit in the deaths of at least forty thousand migrants attempting to cross the Mediterranean to safer shores.[2] While Europe largely responded to the tightening of travel security and counterterrorism efforts of the United States, it also deployed its own version of a racialized spectacle, using discrete, disparate, and rare events to construct an idealized enemy of European values.

Similarly, following the attacks in London on July 7, 2005, the government of the United Kingdom intensified the implementation of its counterterrorism protocol, Prevent. In theory, this sought to address "homegrown terrorism"; in practice, it preemptively targeted largely Muslim community members on the premise that they might spontaneously radicalize.[3] Following the January 2015 Charlie Hebdo attack and the November attack on the Bataclan theater later that year, both in Paris, France saw an escalation in the enforcement of ostensibly secularist policies on Muslim community members; meanwhile news reporting on Muslims surged to an all-time high.[4] All the while only 2.6 percent of terrorist attacks globally have happened in Western countries.[5] Although these events were horrific, the xenophobic construction of an artificially overrepresented and essentialized religious community as a threat in its entirety creates fertile ground for the border and surveillance industrial complex. It also further entrenches technologically what Cedric Robinson calls racial capitalism—that is, the idea that the existence and subjugation of racialized groups has always been central to Western economic systems, and in particular to the development of capitalism.[6]

In addition to terrorist attacks, the so-called Syrian refugee crisis became a justification for the deployment of the full techno-racial wrath of Europe. While smartphones and digital infrastructures were being promoted for use by refugees to help in navigating orientation and to access social, legal, and medical services, these technologies were increasingly used, in the name of security and counterterrorism, to further surveil, detect, and deter migrants seeking safety in Europe. For instance, the biometric technologies that register asylees and extend access to credit are also used in combination with remote-sensing satellite imagery to monitor migration at and beyond Europe's shores.

Beyond Europe's shores, examples include the much-celebrated ID2020 alliance, a partnership spearheaded by Microsoft and Accenture and funded in part by the Rockefeller Foundation to set up a blockchain-based identity system for displaced individuals. It ought to cause some concern that displaced populations were to be serviced by Microsoft, which, as of 2018, enjoys $19.4 million in active contracts with ICE.[7] In other words, the very same infrastructural cloud technologies used for ID2020 to purportedly enable mobility are also used to facilitate the deportation and separation of families in the United States.

In a similar vein, the deployment of IrisGuard's iris scanner technology by the UNHCR and the World Food Program in Jordan's Za'atari and Azraq refugee camps was celebrated as a prime example of how experimental technologies could be used to provide access to credit, in the literal blink of an eye. However, fear from affected communities, aid workers, and academics alike about the technology's invasive and obscure data practices is documented to have disincentivized refugees from registering upon arrival to camps, deterring them from access to critical services.

Companies such as Palantir, a software provider that has powered some of the most aggressive immigration raids in the United States to date,[8] have been awarded wide-ranging contracts with health providers and humanitarian organizations across the globe to manage the COVID-19 pandemic.[9] In exchange for unprecedented access to personal, health, employment, immigration, sexuality, and race data, the company contracted with the UK's National Health Service for a mere $1.3 USD (yes, you read that correctly). This is a sales model Palantir and other surveillance companies (e.g., Clearview AI) use to quickly expand their apparent governmental clientele, allowing them to capture a larger share of the market and hike up the value of contracts later on. In the aftermath of its successful penetration of the realm of public health in the United Kingdom, Palantir returned its attention to its specialty, the reification of hostile migrant environments, as it was awarded a $26 million contract by the UK Cabinet Office for its border flow tool.[10] This reflects a larger trend of pouring hundreds of millions of dollars into militarization and datafication along Europe's borders, a strategy that automates and standardizes xenophobia.[11]

These developments hint at an emergent consensus between technology actors and governments at all levels to share in the management of migrant life. Partnerships spearheaded by technology giants in service of humanitarian and international organizations, backed by massive philanthropic institutions, have experimented with interventions ranging from blockchain-based identity systems to algorithmic resettlement schemes. These violent relationships are, meanwhile, further legitimized and normalized by the narrative of humanitarianism or charity. For every call against the complicity of technology companies in violence against migrants, there is an increasingly generic retort that the company is either stopping human trafficking, finding disappeared children, or feeding the poor and displaced, with minimal data to support such claims.

Mobile populations—communities on the move—face technology-augmented violence in refugee camps, at borders, and long after resettlement. Indeed, such violence is embedded into the very political economy of the West. These are the machinations of the digital periphery and twenty-first-century racial capitalism.

BORDER LEGACIES OF RACIAL CAPITALISM

To understand how we ended up here, we have to go back to 1983, when one of the key thinkers of the Black radical tradition, Cedric Robinson, wrote his pioneering work *Black Marxism*. He problematized and uprooted many of the foundational tenets of Western political thought, particularly around the universal character of capitalism, class struggle, and analysis at large. Robinson provided a critical review and analysis of capitalism, taking Marx and Marxian thinkers to task over their nonexhaustive and rudimentary treatment of race in their analysis of the development and formation of capitalism. Arguing instead that modern conceptions of race were inherent to the exploitation at the core of the capitalist world system, Robinson spoke the quiet part out loud in reminding us that we can only make sense of the manners and methods of subjugation in service of capital through seeing it as *racial* capitalism.

Robinson zoomed in on Marx's analysis of revolutionary struggle in the transition from feudalism to capitalism, noting how former structures of oppression were not in fact negated through this change in a dialectical sense. Rather, racialism survived and was made further sophisticated under capitalism. Robinson's vital contribution here is to point out that capitalism depends on racialized mystifications, which enable those on the higher echelons of it to organize societies in a way that presumes the extraction of value from racialized groups of people as a logical (and at times even divine), normalized, and mundane pursuit.

The expansion of European civilization was indeed catalyzed by the migration of what Romans had referred to as "barbarians," the "North Africans, Italians, Poles [who] cross into Metropolitan France to look for work." This includes peoples of cultures and languages that have since been lost or normalized into Europe's fabric, including "Cornish,

Prusai, Basque, Etruscan, Oscan, and Umbrian." Indeed, assimilation of "barbarians" into the European slave labor force was a critical basis of production deemed necessary for the expanding civilization[12]—a trend that Robinson notes continues to our current day, with slave labor displaced by flexible/zero-hour waged labor and increasingly low-wage gig-economy work. Famously, Britain made use of, and even engaged in the export of, Irish slaves until the arrival and expanding demand for sugar in the 1640s and the subsequent emergence of the large plantation.[13] Racial designations in Europe were marked by several variables, including, but not limited to, those that might define people as "indentured peasants, political outcasts, [the] poor or orphaned females," or any number of other strata belonging to Europe's own "barbarians."[14] Today, similar migration patterns animate the virulent narratives that cast communities on the move into precarity and exploitation.

Remembering once again that race is programmatic, we begin to see the through line between the tropes that first justified indentured servitude, slavery, enclosure, surveillance, and domination, and the forms in which they endure today through violent border regimes. These mystifications have given birth to the digital periphery, which transcends conventional colonial geographies and asymmetries of power; it is a new mystification, fit for the supposedly post-racial era, which nevertheless depends on the existence of underdeveloped, underconnected, undocumented, and undesirable groups for value extraction in service of capital. Below, we trace the various iterations of these mystifications and how they continue to enable the extraction of racial capital at the expense of historically marginalized communities, in particular communities on the move.

MODES OF SUBJUGATION

It is helpful to understand racial capitalism and its contingent practices of bordering as emanating from two key methods of direct and indirect racial subjugation, namely: categorization, that is, the classification of particular groups of people as a category of race; and containment, that is, the entrapment of a particular racial category in either space or perpetual movement for value extraction. These modes have worked hand in hand to commodify racialized people, their bodies, their labor, their spaces, and their movement. They have, through mystifications—that is, tropes, myths, and iconography—innovated atomized and increasingly quotidian manners of value extraction since the early days of capitalism and into the twenty-first century. Race has, in other words, always been the logic at the core of the exploitative function of capitalism, while bordering has made up the means by which those subjected to such logics were captured by the system.

The genealogy of racial categorization is one that moves through media; through lore, numerical representation, images, and iconography; and eventually through computational media. Ann Stoler reminds us both that Robinson's mystifications are part and parcel of the discursive process of racialization and that the concept of race itself is mobile and consistently subject to historical renewal.[15] By thinking of race as programmatic and as a technology in and of itself, we can also see how racial "innovations," as Geraldine Heng describes it, drive many of the imaginaries resulting in everything from normalized and accepted forms of domination and control to mass atrocities and genocide.[16] Contrary to a large body of critical scholarly work on race, including that by Anibal Quijano, who identifies the roots of racial categorization as justification for capitalist expropriation in the discovery of the Americas or the "New World,"[17] Robinson argues that racialism predates capitalism and very much formed the social basis of European civilization.[18] But what were the mystifications used to justify this process of casting out certain groupings of people and exploiting their labor? And how do they endure under logics of technology-augmented racial capitalism today?

Among other things, these mystifications rely on the following:

- *Folklore.* Racism depends on categorical fictions of difference, mythologies, and iconographies, or folklore, to assert hegemony. Communicated through oral storytelling or even visual art, racial tropes are not unique and reoccur in places ranging from thirteenth-century tympana in Rouen to illustrations in English psalters, depicting malevolent executioners, adulterers, and the possessed as having African phenotypes.[19] Karen Fields and Barbara Jeanne Fields explore how the social construction and normalization processes involved with witchcraft are parallel to that of racecraft.[20] For example, racial tropes around Jewish people were so prolific in the twelfth century that they encompassed the scientific, medical, and theological communities, advancing the claim that "Jews differ in nature from the bodies of Western Europeans who were Christian: Jewish bodies gave off a special fetid stench . . . and Jewish men bled uncontrollably from their nether parts . . . like menstruating women. Some authors held that Jewish bodies also came with horns and a tail."[21] Or as per the thirteenth-century encyclopedia of Bartholomaeus Anglicus, whereas white folk are seen to be produced out of colder climates, "hot lands produce black: white being, we are told, a visual marker of inner courage, while the men of Africa, possessing black facets short bodies, and crisp hair are 'cowards of heart' and 'guileful.'"[22] In more recent history, the egregious imagery of the mythical herrenvolk, for whom Europe—and superiority in general—was a birthright, helped create a mythological "rationalisation for the domination, exploitation, and/or extermination of non-'Europeans' (including Slavs and Jews)."[23]
- *Numbers.* One particularly virulent and unambiguously reductive form of categorization emerged through practices of numbering and counting colonial subjects. One of the most well-documented cases of this occurred through the "enumeration strategies" of the British Empire in India where quantification of

colonial subjects was used as a mechanism of atomized social control.[24] The Indian census, in categorizing "the vast ocean of numbers, regarding land, fields, crops, forests, castes, tribes' in the early nineteenth century was pivotal in the creation of "countable abstractions . . . at every imaginable level," propelling forth the deceptive notion that the British were indeed in control of "indigenous reality" as a whole.[25] The mere belief that quantification was useful (indeed even a prerequisite to the survival of world powers) was essential in establishing immutable dependency with the colonizing party carrying out the numbering in the first place, while the actual "significance of these numbers was often either non-existent or self-fulfilling."[26] Undergirding this "colonial *imaginaire*" was a justification for violent efforts to fundamentally flatten and "[clean] up the sleazy, flabby, frail, feminine, obsequious bodies of natives into clean, virile, muscular, moral, and loyal bodies that could be moved into the subjectivities proper to colonialism."[27] Here, numerical categorization is a necessary political tool for establishing the colonial other, providing the moral justification for expansion in the name of civilizing unruly racial subjects. To have recognized colonial subjects as one and the same as the colonizer would have been to afford the same treatment to them, thus raising a moral question around their exploitation, oppression, and exclusion.[28] The prevalence of the use of numbers in categorization goes beyond the Indian census and evokes more recent forms of racialized violence. From 1933, US-based IBM was courting Nazi Germany, eventually providing more than two thousand units of its Hollerith "punch card and card sorting" machines (the early twentieth-century equivalent of sophisticated data management infrastructure), which were used to operate every major concentration camp in Europe, effectively turning individuals into numbers.[29] IBM was also one of many ICT providers to equip the apartheid regime of South Africa.[30]

- *Images*. Throughout the eighteenth to the twentieth centuries, race was (and arguably still is) understood as "a category of modernity."[31] The most important colonized subjects of any given domain are typically the most exploited—and by extension, the most racially profiled and mystified—groups, given their centrality to the colonizer's economic prosperity. One of the most prolific technologies used to generate a spectrum of modernity mapped onto race has been the use of photographic images. It has been well established that humanitarian photographs, taken from missionary expeditions from the 1840s onward in colonial contexts were central to the eugenics and scientific racism movements.[32] They served as an illusion of irrefutable difference between subject and savior, making synonymous the suffer*er* with suffer*ing*.

The use of humanitarian photography by organizations and religious actors from the eighteenth century onward became the medium through which "distant spectatorship" was made possible, while at the same time relativizing pain and suffering in new ways. As the industrial revolution paved the way for distancing a people from the drudgery of heavy work, the suffering of "distant others [became] more proximate."[33] Ironically, as technological and medical advancement reduced pain in some populations, the same advancements increased the volume and magnitude of violence (e.g., as a result of war) elsewhere; a new "moral posture" emerges—following the wide dissemination of stories of distant suffering—that paints suffering, and by extension those who suffer, as fundamentally distasteful.[34]

Photography, according to Ruha Benjamin, cemented visual cues of racially stratified difference and carried with it the illusion of objectivity and neutrality, in stark contrast to medieval art's depictions of race (as discussed above).[35] Today, processes of categorization happen everywhere from data-labeling online content for machine legibility and extracting biomarkers from facial features for facial recognition surveillance, to categorizing communities on the move arriving at the border as irregular, undocumented, criminal, or worse.

While categorization allows us to see how subjects of borders are curated and constructed along racial lines, practices of containment have been deployed to keep those attributed to particular racial categories contained in space, or in perpetual movement, for their exploitation. In England, in a bid to assure greater socioeconomic fortunes, "enclosures, the poor laws, debtors' prisons [and] transportation (forced emigration)" were used liberally to contain racialized communities.[36] This was a critical component of both settler and internal colonialism, which saw the systematic erasure of indigenous people, land, objects, and reality altogether, "cleaning" up what was considered wild and untamed space for conquerors, and transforming them into commodities intelligible by whiteness—or what Nicholas Mirzoeff refers to as the violent construction of "white space."[37]

Following the increasingly lucrative sugar industry, and in a bid to extend its production capacity, England further intensified its use of slave labor—and African labor, in particular. African slaves were peripheralized through their emergence into the European invention of the "Negro"[38] while being kept in place through servitude on plantations, or in perpetual movement as labor commodities on the slave market, transferred from one port and plantation to the next. Practices of containment follow on geographies resulting from practices of categorization; once racialized, populations can be cordoned off in service of their exploitation and the extraction of racial capital.[39]

Today, the continued function of "containment" has been masked by largely universalized and taken-for-granted notions of borders as features of the Western liberal model of nation-states. Harsha Walia's pivotal work points to "border imperialism" as a lens through which we can analyze how communities on the move—and their bodies in particular—are instrumentalized by nation-states, imperial aspirations, and the security imperatives of global racial capitalism.[40] Importantly, while borders are conventionally thought of as the physical demarcation of space, border imperialism includes the "conceptual borders that keep us separated from one another."[41] Understanding containment as a function of border imperialism helps us make sense of the ways in which borders continue to have salience, not just before and during refuge, but also after resettlement in the everyday lives of displaced populations.

Containment could, as discussed above, emerge through extreme mobility, referring specifically to "the ways in which migrants'" movements and presence are troubled, subjected to convoluted or hectic movements and protracted moments of strandedness," as Martina Tazzioli argues.[42] These policies are necessarily as much about spatiality as they are about temporality. Thus, containment strategies are not necessarily reliant on a physical carceral space that enforces detention, but could consist of a mixture of policies, including those that render conditions of displacement hard to escape, make it impossible to reach intended destinations, or even disrupt how information about the process of immigration is obtained.[43]

This helps us arrive at the point that the technical machinations of borders today are fundamentally elastic, increasingly fused with bodies on the move, looking to foreclose any evasion long before and after an individual arrives at a material frontier. They seek to lock migrants into a prescribed identity and/or derelict conditions of labor for exploitation and political capital.

CATEGORIZING AND CONTAINING THE UNRULY MIGRANT

From its formation through to its current iteration, racial capitalism has been contingent on the utility of these two aforementioned modes of subjugation to capture the subjects and sources central to the production, accumulation, and protection of wealth. Building on Paul Gilroy's theorization of the Black Atlantic,[44] the migration scholar Alessandra di Maio refers to the process by which these geographies of violence are created—through which the possibility for modern Europe arises in the first place—as the "Black Mediterranean."[45] The Black Mediterranean is constructed internally (e.g., through racialized integration politics)[46] as much as it is externally (e.g., through border imperialism).[47] States and nonstate actors engender the Black Mediterranean through establishing an existential (e.g., terrorism) or socioeconomic (e.g., unemployment, austerity, and staggering wages) threat to an idealized notion of European civilization.

This state of migratory exception is utilized as a political technology in response to manufactured political events (e.g., in the face of forthcoming elections or other global trends related to geopolitical interests), intended to reorder and rally the body politic around the idea of the cohesive nation-state.[48] In the words of Robin D. G. Kelley, "The Black Mediterranean is about the fabrication of Europe as a discrete, racially pure entity solely responsible for modernity, on the one hand, and the fabrication of the Negro, on the other."[49] One stark example is the makeshift refugee camps set up on the Greek "hotspot islands" facing Turkey, following the EU–Turkey deal of 2016.[50] Here, said hotspots served as temporary containment sites for migrants, part of a strategy of usurping efforts of reaching mainland cities like Athens.[51] Yet, the experiences of strandedness, coupled with invasive identification procedures (e.g., through mandatory biometric registration procedures), have been documented to have compelled a greater imperative for migrants to move.[52]

The EU spent much of the years that followed the Syrian crisis of 2015 espousing a two-pronged public narrative that foregrounded its generosity toward migrants while maintaining a strict position on "stopping unruly mobility" and "keeping migration across Europe to the minimum." It also exercised its governmental logic through keeping migrants "on the move," forcing them "to undertake more and more erratic and diverted journeys, as a result of the many internal transfers they are targeted for."[53]

This in turn predetermines the exploitation of migrant precarity, by generating a hostile environment under which only limited and marginal forms of labor are possible while others are criminalized. All of this while states derive political capital from claims made on the back of the very existence of such communities; companies derive profits from the development, deployment, and reinforcement of state-sanctioned technologies, and research institutions win massive grants from European agencies for the development of so-called solutions to the refugee crisis. The trade in the bodies, images, subjectivities, and environments of communities on the move is no small venture; between 2014 and 2020, the European Border and Coast Guard Agency—known as Frontex—invested €434 million on surveillance and IT infrastructure. For the years 2021–2027, the European Commission has earmarked some €34.9 billion for border control.[54]

The artificial construction of the undocumented, ungrateful, and in other ways irregular migrant, named so for political expediency, is part and parcel of what Walia calls the process of "illegalization," through which the neoliberal "conditions of permanent precarity" are created and maintained, while at the same time "legaliz[ing] the trade in their bodies and labor for domestic capital."[55]

Through this charitable-cum-disciplinary dualism, Europe has forged conditions of absolute transience for migrants, condemning their fates to protracted conditions of movement without resolve. Europe manufactures lucrative states of migratory exception that maintain and reinforce the Black Mediterranean by universalizing migrants' transience and "keeping them in transit." Through practices of categorization and containment, colonial powers—and Europe in particular—have laid the groundwork for an architecture of racialized violence and capital extraction that continues to animate racial capitalism and its digital character to this day.

2

THE MAKING OF THE DIGITAL PERIPHERY

THROUGHOUT ITS EVOLUTION, from formation into the digital age, racial capitalism has relied on categorization and containment. Through categorizing colonial subjects (e.g., as per Britain's deployment of the Indian census), an illusion of control and ownership can be exerted on racialized communities. Through containing (or enclosing) categorized communities in space to work land as indentured servants or extorted laborers, value can be extracted and turned into profit. The datafication of displacement—and indeed of racialized communities in marginalized contexts in general—over the last two decades has replicated and augmented these modes. Individuals and communities on the move are now categorized under the flattened digitalized categories of refugee, Black, development, big data, and third world, and their states of immobility (being confined to a camp, detained at detention site, unable to move outside particular areas out of fear of persecution and deportation, or even being kept in perpetual motion) exploited for the deployment of experimental technology interventions. This is the digital periphery, where technology companies situate themselves as political stakeholders in migration governance to gain new ground for technology deployments. The digital periphery is mobilized in three major ways by those who benefit from it: through techno-development, techno-space, and techno-government, which can be thought of as three distinct forms of bordering. It operates techno-developmentally through the instrumentalization of modernization tropes by tech actors, inherited from a long tradition of colonial representations of the needy, particularly in the humanitarian and development sectors. In techno-spatial terms, city-dwelling migrant communities are conditioned for persistent precarity through the threat of surveillance, which forcibly contains them in physical and virtual space for the continued deployment of tech on them. Finally, the digital periphery matters techno-governmentally, as governments and institutions rely on technology actors and their digital urban interventions, symbiotically, to identify, exploit, and control migrant communities.

TECHNO-DEVELOPMENT

Techno-development is a way for us to make sense of how technology actors draw on aged modernization tropes and neocolonial international development discourse to legitimate their intervention in the lives of others. These racialized and yet depoliticized and ahistorical narratives situate global majority countries and the postcolony as in need of so-called Western innovation to emerge from the depravity of their "backwardness."[1] Take the example of a mobile app or smart ID system intended to improve the lives of refugees. Under this banner, tech companies flatten and essentialize an entire grouping of diverse people with unique circumstantial factors into one group, usually with accompanying images of Black and Brown people in camps, borders, and other humanitarian contexts. This digital manifestation of racialized refugees is sandboxed through the particular technology in question, allowing technology actors to justify their continued investment in experimental technologies with at best, unknown impacts, and at worst, devastating consequences for the lives of communities on the move. Flashy projects such as ID2020,[2] IrisGuard,[3] and more recently Worldcoin[4] build techno-utopian fantasies on the backs of communities on the move. They treat precarious conditions of displacement as either inherent features waiting to be alleviated by the provision of blockchain-based biometric identification systems or as a moralizing rationale for why the technology must exist, even if no one asked for it. Border violence was, however, never a technical problem mitigated by multifarious identification experiments, but rather a predominant feature of the austere and xenophobic politics of the modern nation-state.

These interventions in migration governance enable tech actors to extract the raw materials needed to develop further products (e.g., biometric, statistical, and behavioral data).[5] They also drive financial capital from philanthropic organizations, venture capitalists, humanitarian actors, and governments to the tech sector, which intervenes on their behalf. The public fantasy that perceives those in precarious positions based on race, ethnicity, nationality, and/or religious affiliation as burdens or threats is in fact what generates the self-perpetuating rationale for investment in sophisticated technologies to keep them at bay.

In other words, actors with stakes in the tech world veering into the world of politics, crisis, and humanitarian management is no longer news. At the height of the so-called Syrian refugee crisis, the organization known as Techfugees was launched by *TechCrunch*'s editor-at-large Mike Butcher. In a *Verge* article titled "Hacking the Refugee Crisis" from 2016, Butcher lays out his plans for the organization to be the great synthesizer, bringing "the tech community together, at least in Europe, to address [the refugee crisis] in the ways I know they are capable of."[6] He describes how areas ranging from housing, integration, and education will be tackled through technical expertise, delivered at the different chapters of Techfugees. The organization espouses a techno-solutionist modernization narrative,[7] purporting to exist "to empower the displaced with technology," with "many technology businesses and entrepreneurs [having] expressed an interest in helping non-profit organisations to automate their processes."[8] Again, one would be forgiven for thinking this narrative was directly lifted from the international development sector, where colonial–missionary vernaculars are alive and well. If only we could give poor refugees the right technology, surely their material and political circumstances of refuge would be surmounted, right?

In 2016, the organization became the meeting place for agencies like UNHCR, UNICEF, and the Red Cross, as well tech companies, including Google, which continued to sponsor Techfugees until at least 2019. Techfugees positioned itself as a mediator in the "crisis" and as a spokesperson for refugee needs. In an event hosted at Google HQ in Denmark (ironically the state with some of northern Europe's toughest asylum, immigration, and integration policies), the Copenhagen branch of Techfugees aimed to "[hack] social inclusion for and with displaced people,"[9] framing a question of political will in terms of technical capacity. The organization is only a symptom of greater transformations in how migration governance and border violence came to be discursively depoliticized and banished to the realm of do-gooder entrepreneurial development initiatives. Even states with the most prohibitive immigration policies are rushing to announce their latest technology "for good" initiative that would revolutionize the crisis. In 2017, French president Emmanuel Macron announced a partnership with Airbnb's recent Open Homes initiative to match refugees with housing. This made France the sixth nation to do so, in addition to the United States, Canada, Germany, Greece, Italy, and Spain.[10]

Techno-development has figured into technology discourse for decades, most prominently after the collapse of the discursive dualism between real space and cyberspace. In part a result of the role of social media platforms in the Arab Spring, technology actors increasingly came to situate themselves more outwardly as operators of political space. To understand the precursor to the convergence between the technology sector and humanitarian/international development discourse, we must consider two major developments: first, changes Anita Say Chan documented in her observations of the FLOSS community and their adoption of explicit political agendas;[11] and second, the accelerated growth of Silicon Valley technology corporations and their co-option of the FLOSS community.[12]

FLOSS stands for free / libre / open source software. Chan documented how the "highly skilled information class" that made up the FLOSS community came to be at the center of a political awakening in the tech world: "If geek culture and hacker practices had once appeared abstracted, obscure, and separate from broader social concerns, by the turn of the century such social distinctions no longer seemed to hold."[13] The activities carried out by the FLOSS community became a battleground for Western liberal norm diffusion, with the "the right of free speech, assembly, petition, and a free press, . . . the stability of property and especially IP law," taking center-stage in the resolve of hacker communities,[14] not too differently from how hacker communities today are turning their attention to migration. Clashes between government and global hacker groups and their politics became commonplace with the emergence of the Anonymous hacker collective. Hacker communities expanded their focus from advocating for free and open software toward the "'real space' realm of established politics."[15] In other words, the emergence of the political hacker came to increasingly represent an alternative to politics as usual, a potentially fundamental shift in power relations in technical terms.

Concurrently, the emergent big tech companies at the time (notably the GAFAM five of Google, Apple, Facebook, Amazon, and Microsoft) underwent a "dramatic self-transformation" in an effort "to maintain dominance." A prominent example of this is Facebook (now Meta), which not only sought to outwardly flatten the hierarchy between corporations, users, and Facebook itself,[16] but also positioned itself as a natural extension of the FLOSS community as we know it and adopted a political identity.

Facebook was only one among a handful of tech giants whose libertarian inclinations paved their way to the global stage not only in the tech industry but also in the world of activism and politics. Wael Ghonim's notable phrase in the context of the removal of Egypt's Hosni Mubarak reverberates through every technology actor's insistence that their product will be the one to democratize and empower the oppressed: "If you want to liberate a society, just give them the 'Internet.'"[17] Silicon Valley giants have since ventured far and wide, often with one foot in humanitarian affairs and another in development. Beyond Facebook's Internet.org and Libra initiatives, Google has launched connectivity projects ranging from free Wi-Fi in and around refugee camps (see its Signpost and Refugee.info projects) and beaming service to rural areas using weather balloons via its sister company Loon, to installing massive subsea-level internet cables connecting Europe and southern Africa. Furthermore, Sidewalk Labs, owned by Google's parent company Alphabet, heralded smart cities as the natural evolution of urban environments, to bring in greater efficiency and safety. Meanwhile, Palantir, the controversial tech company that aided ICE in its deportation raids of undocumented immigrants in the United States, entered a partnership with the World Food Program.

TECHNO-SPACE

Over the last decade, cities have been the site of a battle over techno-spatial real estate, in everything from the internet of things (connecting everyday infrastructures to cloud-controlled processes, including transit entry, connected sensors, and AI-driven surveillance and early-warning systems), the process of generating "digital twins" of cities (through digitalization, further datafication, and atomization of city life and processes),[18] and blockchain-based identity and e-citizen services purporting to generate greater transparency and control among citizens.[19] The pursuit of the smart city has given governments the ability to erect a mirage of order and control through technology deployments while allowing technology corporations to become an essential part of the performance of governance, often without being openly acknowledged as such.

At its extreme, cities have been weaponized for speculative and, at times, pseudo-scientific experimentation with remote biometric recognition tools such as race, gender, and emotion recognition, exacerbating processes by which minority communities are further targeted, surveilled, and persecuted.[20] Elsewhere, cities are being built with unprecedented data-harvesting at their core, threatening basic freedoms, automating repression, mass, arbitrary, and targeted surveillance, and shrinking civic space.[21]

In cities such as these, the entanglement of digital urban infrastructures with corporate power has been increasingly commonplace. In New York, for instance, the Mayor's Office appointed tech giants from Meta, Google, IBM, Microsoft, and Verizon to its NYCx Advisory Board and several other elite city committees.[22] This team-up of tech giants served to ensure NYC achieved its mission of becoming a "testbed for new technologies [transforming] the relationship between city government, community, and the tech industry."[23]

Cities are quickly becoming construction sites for data pipelines, cultivated, expanded, and fed by the movement of lives of the most vulnerable. More than ever, Silicon Valley giants are hollowing out everyday life in our cities, in service of smart-fetishistic futures and fantasies. Imagine, for instance, how the process of navigating the city via augmented reality might be tailored to what the city knows about you in particular. What might you be barred from accessing? How might you be rerouted, subject to your wealth, race, immigration status, or criminal record?

These global developments are especially worrying given the situatedness of cities as sanctuaries, historically providing scale and numbers for persecuted communities to hide under the anonymity of the sheer vastness of urban environments. They also have implications for the contemporary and future distribution of power within cities, between the privileged and those at the margins. The groundwork has already been laid for the emergence of the "mobile outcast ghetto" where underprivileged and historically marginalized communities are increasingly ending up. Here, "logistics orchestrates the control and management of surplus populations, keeping them in their (social and economic) place, even as they move about the city."[24]

By way of an example, "The 'bad part of town' will be full of algorithms that shuffle you straight from high school detention into the prison system. The rich part of town will get mirror-glassed limos that breeze through the smart red lights to seamlessly deliver the aristocracy from curb into penthouse."[25] In such future cities, our machine legibility—and conformity—plays a huge role in our ability to access basic rights and services. Your daughter may not be able to get her subsidized medication, owing to what might be thought of as data debt: the decision not to share certain data with a party whose advertising profit (made at your expense) in part funds medical care. Your human dignity and access to rights, in other words, will be at the whim of your contribution to capitalism.

It is not inconceivable that, in this future city, our rights depend on aggressively data- and biodata-intensive human–computer interaction—biometric interfaces that require reading vital signs and other health indicators constantly and passively, with the insurance costs of those with the poorest health (or most strenuous working conditions) rising in line with such metrics. Fitness trackers and smartwatches have already laid the groundwork for this, with facial recognition–enabled smartwatches even being proposed as a form of incarceration for migrants with criminal records in the United Kingdom.[26] In such scenarios, we are penalized for opting out of the cyborg future; those of us unwilling to make ourselves known must be prepared to live in conditions of permanent turnstile-jumping (metaphorically and, possibly, literally).

These developments, especially in cities—where the pervasiveness of cashless transactions is a de facto expansion of digital urban infrastructures that control how we engage with goods and services (and indeed *who* can engage with these)—may also lead to a ban on cash, eroding informal markets and safe spaces for those living in fear of surveillance, persecution, and expulsion. In a particularly dire speculative scenario, thoughtcrimes[27]—detected through a mixture of responses to high-frequency advertisement techniques, attention-trading,[28] and biometric markers based on debunked ideas around microexpressions and phrenology—may lead to preemptive arrests, detention or expulsion, in attempts to flatten society to an ideologically homogenous whole.[29]

By containing communities in virtual and/or real space, digital urban infrastructures work to border and exploit communities. Layered on top of existing racialized peripheral geographies of violence,[30] people marginalized by borders are identified and layered with a digitally mediated symbolic enclosure[31] in the digital periphery. This allows for a more granular execution of border functions, beyond conventional Global North / South divides. Every time an app claims to be helping refugees through fostering their entrepreneurship, it is also telling us that it would prefer that said user be more self-reliant, engage in more precarious and individualized forms of labor, independently of the state. When a person is asked to sign up for a biometrically enabled digital ID, they are being asked to give up the possibility of invisibility; in exchange for your daily bread, you must make yourself fully known to authority (or risk being perceived as "illegal").[32] Equally, when we are convinced that social justice can be accomplished through engineering activities, the possibility for dissent and contestation is eroded. The "New Jim Code," as Ruha Benjamin writes, is further strengthening racist structures, which "not only marginalize but also forcibly center and surveil racialized groups that are 'trapped between regimes of invisibility and spectacular hypervisibility,' threatened by inclusion in science and technology as objects of inquiry."[33] Foregrounding technologically mediated marginality in racial capitalism opens the possibility that even the most well-intentioned interventions are in fact about spatial capture on the backs of peripheralized subjects.

TECHNO-GOVERNMENT

Governments and institutions rely on technology companies, symbiotically, to identity and control otherized communities. Double-edged digital infrastructures, which serve as both services and surveillance, are a means by which authorities situate surveillant and exploitative systems as public goods. Communities are in turn made dependent on such systems for access to a bare minimum of information, services, and their cities at large. While companies that provide the raw materials through which governments can further assert their power of subjugation thrive under these conditions, the rest of us lose.

In the digital age, technologists increasingly seek to encode documentation into the human body by introducing near-immutable forms of stratification that encode marginality. This recasts the stateless and those with precarious immigration status as digital others. These austere biometric projects accentuate and reproduce xenophobic state logics, which treat communities on the move as suspicious by default, modern vagrants who deserve only the bare minimum. As recently as July 2023, Worldcoin, the cryptocurrency project founded by OpenAI CEO Sam Altman, announced its plans to provide a zero-trust decentralized human identification system, named World ID, aimed at distinguishing humans from bots.[34] Through the rollout of some 1,500 camera-equipped "orbs" across over thirty-five cities, Altman promises a roughly fifty-dollar equivalent of its Worldcoin currency in exchange for scanning your eyeballs and registering them with a unique identification. It wasn't long before World ID was being touted as a natural intervention in refugee camps;[35]soon after the orbs' debut, OpenID announced that it would allow governments and companies to use the World ID system as well.[36]

World ID is only the latest in a string of corporate initiatives that work with and/or entice governments with greater degrees of control over communities on the move. Refugees, as the organizer from Mijente reminded us previously, are at the early stage of the technology production innovation curve. In the year following the Syrian refugee crisis of 2015, Europe saw a meteoric rise in the number of digital urban initiatives purporting to empower refugees.[37] This came following nearly two decades of investments in high-tech smart border systems in the aftermath of 9/11,[38] and at least a decade of techno-humanitarianism in camp contexts and along routes of transit.[39] Digital technologies have even served as the symbolic front for the race to the bottom on stricter immigration policy in Europe,[40] a political strategy through which governments can claim that they are proactive in their management of borders.[41] The increased digital presence of refugees, and migrant populations in general, has made it easier to track refugees by a range of interested actors, including keenly interested academics, NGOs, and rescue coordinators, to more overtly sinister actors, including security agents who intercept migrants between borders and/or deportation agents (including but not limited to ICE).

Researchers have been critical of the adoption of digitized borders by governments in Europe, designed to help preempt mobility. From mathematical formulas that neatly calculate the numbers of refugees any given European country should receive (based on variables including the number of asylum applications to the country, unemployment rate, and GDP),[42] to technical innovations that surveil mobility,[43] there is no shortage of examples of how techno-governmental entanglements have shaped and determined refugee destinies. Public and institutional imaginations have remained captivated by technical solutions to the refugee problem, starting from arguably the most prominent of migration-related international agencies, the UNHCR, which houses its own Innovation Service.[44] A central logic underpinning its existence is "self-reliance" as an issue central to the plight of the refugee. With bodies like the UNHCR advocating for private-sector solutions to what is largely a problem of political will—placing the onus on individual refugees to "improve their own lives"[45]—it is no surprise that technology corporations have become significant actors in mobility governance.[46]

These initiatives, in turn, have potentially devastating consequences for those on the move. Helle Stenum, for instance, writes on the difficulty of benefiting from strategies such as "flexible identities" and "de-identification," used predominantly to circumvent deportation orders or facilitate safer transit.[47] When documentation of personal identification is encoded biometrically, through blockchain-based digital ID systems, the movement and fugitivity of refugees is foreclosed, tightening the state's hold on the individual in question. For the displaced, the ability to navigate visibility and invisibility can be a matter of life and death; the body borders generated by biometric initiatives increasingly close in on the possibility for resistance[48] by situating refugees under conditions of hypervisibility.

The surveillance and control of communities on the move is surrounded by techno-governmental entanglements that are moving closer to our cities with each passing moment. Here, digital technologies (whether through e-passports, facial recognition, or biometrics) increasingly determine whether individuals should be allowed entry through particular borders, or whether they should be detained and deported; whether they can access goods and services in refugee camps, or whether they will live precariously; whether they will be able self-determine, or whether they will be policed into conformity;

whether they will be granted rights conditionally, or rendered digitally determined pariahs. While these dynamics are very much inherited from the birth and logics of the modern passport system in the early twentieth century,[49] techno-governmental entanglements bring borders even closer to bodies.

As part 2 demonstrates, practices of categorization and containment weave together places as disparate as refugee camps, widely established as being rife with apps as well as digital and biometric ID and surveillance systems; and cities, where the effects of digital urban infrastructures on marginalized migrant populations remains murky. The transformation of racialized communities on the move in New York and Berlin into the digital periphery not only sustains technological experimentalism, data extraction, and the mobilization of vast amounts of capital for profit, but importantly drives symbolic and political capital from governments and humanitarian actors to tech actors, who are, consequently, increasingly engaged in exercising governance.

PART TWO

RACE IN THE DIGITAL CITY

3

XENOPHOBIC ROOTS, TOLERANT FACADES

FROM ELLIS ISLAND IN NEW YORK, as the romanticized ideal of sanctuary, to *Willkommenskultur* purporting to hail an era of unprecedented hospitality for refugees in Berlin, these historical cities of refuge have been the locus for interplays between xenophobia and liberal performances of tolerance. These narratives have been instrumentalized for profit, prestige, and control, in the name of communities on the move. In this chapter, I trace how foreclosures on migration and citizenship are rearticulated and disaggregated over time, creating the conditions for the emergence of the digital periphery, which mask and reinforce historically salient notions of otherness and race. In New York, I unravel how often brutal, eugenicist, and violent processes of assessing arrivals into the port in the twentieth century were subsequently revamped and disaggregated into everyday urban life, paving the way for the very digital urban infrastructures that are bringing borders closer to our cities. In Berlin, I look at how the shift from "foreigners' law'" to "immigration law" was motivated by an aging workforce and the inextricable relationship between the tech sector and immigration policy. A picture emerges of the relationship between capital, borders, technology, and governance.

INSTITUTIONAL DECAY IN THE SANCTUARY OF NEW YORK CITY

New York is one out of twenty cities with over one million foreign-born residents.[1] It is counted as one of the most "hyperdiverse" cities, next to cities such as Toronto, under the following criteria: "At least 9.5 percent of the total population is foreign born (this is the average percent of foreign-born stock for developed countries according to the United Nations)," "no one country of origin accounts for 25 percent or more of the immigrant stock," and "immigrants come from all regions of the world."[2] Moreover, New York's rich history of immigration sets it apart from other hyperdiverse cities. In many respects, New York is the polar opposite of many European cities—even in the seventeenth century, under Dutch rule, purportedly "18 languages were spoken in the streets."[3] Following the Civil War, there has been steady increases in immigration to the city. Between 1790 and 1860 alone, the population rose from 33,131 to 813,669; by 1950, the population had grown to 7.9 million. While this is attributable partly to the "declining transatlantic transportation costs," it is the "immigrant-specific social and political infrastructure that made, and continues to make, New York a magnet for immigration": "large communities of immigrants from specific countries [that] allowed new immigrants to come to New York while continuing to speak their own language [and] suppliers provided commodities that were closer to those . . . consumed in their home countries."[4]

New York City's immigrants—and their subsequent descendants—have historically played "the role of 'hosts' to . . . new arrivals, passing on lessons about New York and the United States, and shaping newcomers' thinking and actions."[5] Anchor communities are in many ways the reason for the city's success in attracting and keeping newcomers across its five boroughs (Brooklyn, Queens, Manhattan, the Bronx, and Staten Island). That is not to paint a romanticized image of the lives of the city's immigrant populations, who continue to live in a fraught state. Structural marginalization across racial lines looms large—Black and South Asian communities continue to be subject of heavy surveillance, and deportations of Pakistani residents following 9/11, and now Latinx communities in the wake of the Trump administration's border wall, have been on a steady rise.

As chapter 4 will explore, the transformation of New York into a site of digital migration control, and the transmutation of its vulnerable migrant populations into the digital periphery, has been possible owing largely to the neoliberal tactics deployed by the federal government. Targeted institutional decay has been an important means by which the city as a sanctuary has been threatened. The Trump administration's incessant legal battles to maintain its regime of immigrant violence—including but not limited to family separations, detention under inhumane circumstances, and deportations—while significant, were not always successful. To some extent, they have served as distractions from less obvious policy changes, with devastating consequences for institutions key to immigrant survival in the city. At the core of this is the systematic and deliberate obstruction, and subsequent defunding, of voluntary agencies (VolAgs) in charge of providing reception and placement (R&P) services in New York.

On January 28, 2017, President Trump signed Executive Order 13769, otherwise known as the "Muslim ban," prohibiting individuals, including refugees,[6] from the Muslim-majority countries of Iran, Iraq, Libya, Somalia, Sudan, Syria, and Yemen from entering the United States for ninety days. The order also brought an effective suspension to the refugee resettlement program for 120 days, while slashing the determination for annual refugee intakes by more than 50 percent.

In the fall of 2018, I was at the offices of a refugee resettlement organization (RRO) in Flatbush, Brooklyn. Caseworkers were chaotically scrambling to find avenues for affordable housing for refugees, in addition to searching for supporting evidence of their capacity to resettle refugees. Not only had the Department of State issued strict accommodation guidelines that had to be met for a housing option to be considered legitimate (requirements that were all but impossible in New York with the funding allocated per client), but had also determined that the US Committee for Refugees and Immigrants (USCRI) would be receiving a significant cut in funding, owing to its inability to house their targeted number of refugees.[7] The reason for the decreasing number of resettlements? The so-called Muslim ban, which was stifling the number of refugees able to enter the United States. USCRI's grant loss, in turn, had a compounding effect on the RRO I was embedded in at the time, Reset, which experienced the shrinking of sanctuary space firsthand. Without funds for R&P, Reset would no longer be eligible to provide resettlement services. It faced the prospect of not only having to turn away clients but also having to potentially manage an increasing number of asylum-seekers and undocumented immigrants, whose safety could not be guaranteed. Rather than an unfortunate bureaucratic mishap, this appeared to me an acute example of targeted institutional dismantlement.

In many ways, the US refugee program has always been in a state of dismantlement, created for the purpose of being destroyed. This is where my engagements in Brooklyn began. In Flatbush, the pathways of Caribbean, Black, and foreign-born immigrants who migrated to the area in the 1980s cross.[8] Here, you will encounter "the largest concentration of undocumented immigrants" in Brooklyn.[9] Unsurprisingly, it is also home to one of the largest RROs in the borough. Reset was founded in the 1970s and began economic development work in immigrant communities shortly after, just as the area was at the height of a major demographic transformation.[10]

I started working at Reset as an in-house researcher and volunteer during a time when they were being assessed for funding. The devastating impact of Trump's Muslim ban meant that New York— among a host of other sanctuary cities—had been identified as an "area of concern" by the US Department of State (DOS). This, I am told by a managing caseworker, Anita, is code for failing to deliver on resettlement targets while at the same time being considered hotspots for refugees and other vulnerable immigrant populations. "After September 11, nobody arrived," Anita tells me. Before that, at least six hundred refugees had arrived from Kosovo in 1999. Applications were quickly approved as airplanes were filled and flown into New York, with RROs and, somewhat surprisingly, the military working together to check people in. "The government has historically created good programs, but the administration—in its failure to succeed in courts [in curbing immigration]—challenges resettlement through altering and adding rules to the program."[11] Anita explains that there's a requirement to resettle one hundred refugees over the financial year to qualify as a service-providing VolAg for R&P. Consequently, this is filtered down to the USCRI's partnering RROs, who, due to the pressure faced by their VolAg, are forced to compete with one another, whereas there had previously been cooperation (e.g., between Reset and other VolAgs, such as the International Rescue Committee).

It is clear that the defunding and decrease of refugee service infrastructure is deliberate, with the administration having determined twenty-six cities as areas of concern, New York being one of them. Dino, Anita's deputy—himself a former refugee from Bosnia—notes that they have few, if any, tools at their disposal to fight Trump's "rising invisible wall."[12] "His [physical] wall is getting smaller. So, this is his contraction strategy"—in other words, his wall was a distraction shrouded in layers of more subtle transgressions. The ongoing reduction of some three hundred RROs to merely 175 at the time, combined with the administrative hurdles of strict housing requirements and more invasive screening measures, including new technologies like DNA testing, transforms individuals fleeing in search of a better life to—in Anita's words—"pipeline people." As pipeline people, refugees, asylees, and undocumented immigrants alike become de facto "foreign policy escape-valve[s],"[13] kept in political suspension through their permanent containment in either the bureaucratic process or through perpetual movement to escape said process. As this chapter attempts to show, targeted institutional dismantlement, then, involves the material and symbolic capture of fugitive subjects and communities on the move.

In anticipation of these developments, Reset had found itself preparing by turning its focus away from refugees and toward asylees. While fewer people were able to come through via the refugee program, an increased number of individuals were in turn forced to find alternative pathways of refuge. While not strictly in compliance with what constitutes R&P services, Reset had to figure out a way of tending to this increasing population: "So that's, that's the challenge. It's a very difficult question, I mean . . . hard to answer . . . the reason is if you come to the US airports or US port . . . and you claim asylum . . . you will be put in detention," says Yunus, a caseworker I came to work with closely.[14] Originally from Rwanda, Yunus had made it to the United States—Atlanta specifically—just before the window closed in 2000: "Yes, I came as a refugee through the IRC [International Rescue Committee]. I came in 2000, I started working with them right away." He explains that the uncertainty around the availability of jobs, as well as gaps in his experience, led him to apply for the position of Arabic assistant with the IRC. "They were going to resettle a huge number of Iraqi refugees, and they offered me a job. I didn't have any other job and didn't have any experience; when I left I was coming straight from school so . . . I'll take it! And from there I became a case manager." When Yunus first arrived in the United States, he was placed within the Burundi community for housing; with few Rwandans in Atlanta at the time, it was important to be close to an anchor community who spoke the same language. Yunus explains that despite the difficulty, they were obligated to house him: "As a free case,[15] they . . . have to find you housing with furnishing, food, everything. So, for the free cases, you mostly depend, for everything, on an agency. For a US tie case,[16] on the other hand, the agency provides only basic services, the 'tie' provides housing and some transportation."

Yunus worked at IRC in Atlanta for four years before moving to New York. While we have long dispelled the myth that there was ever an American golden age for immigrants (especially the undesirable kind), Yunus and some of his clients were part of the 69,886 refugees admitted in the year 2000 before admissions were halved the subsequent year. Even more significantly, refugee admissions had peaked at 122,066 in 1990 before dropping steadily every year, totaling a meager 22,491 admissions by the start of this research in 2018. In some respects, Yunus and his contemporaries had arrived just before the window slammed shut. They had escaped the increasingly hard borders of the hostile immigration environment while finding themselves subject to increasingly invisible ones.

Others were not so lucky. During one of my "client days" at Reset (designated days for clients picking up checks and having appointments with their caseworkers), I met with Abdo, a Darfurian refugee who had spent the last eight months in one of ICE's seven notorious detention facilities in California. Certain that his application to participate in the refugee resettlement program would be rejected, Abdo had no other choice but to find an alternative means of escaping to New York, which almost certainly involved ICE. "I wish there was more focus on detention centers. Detention centers and language learning," he tells me. Even though Abdo had arrived with a companion, his friend had soon been released; Abdo reiterates that knowing the language was crucial in minimizing the barriers faced by pro bono lawyers and ICE agents, an advantage he did not benefit from at the time.[17] In his experience, knowing or not knowing English makes the difference between being in detention for one versus eight months. Abdo and many others in his situation had experienced immigration status-related turmoil for linguistic reasons, despite the fact that the country does not officially enforce a national language. ICE did not make any concessions for these circumstances.

Acknowledging Reset's place in this impossible landscape, Anita catches me after my conversation with Abdo, stressing the authorities' lack of interest in socioeconomic success as a measure of successful immigration and integration: "It's defined by interaction with bureaucracy—you're only integrable if you are yet to integrate. If you succeed, you're no longer integrable, so you're no longer useful." Anita portrays the immigration process as fundamentally concerned with capturing, in one form or another, migrant populations, with no endpoint in sight: "Integration, in most cases, is performing the role of an *irregular*—of a transient presence in the populace."[18]

Abdo's story is not unique. Over the course of my four months fieldwork in the city, I speak to nearly a dozen asylees who report similar dilemmas. In general, refugees, asylees, and individuals in vulnerable immigration conditions are faced with a Hobson's choice:

Choice 1. Apply for the refugee resettlement program or seek asylum upon arrival at a US port of entry. Either way, the individual has to surrender agency to the disciplining function of the bureaucratic process of the refugee "pipeline." Here, the migrant exists as a public charge in the eyes of the state.

Choice 2. At the expense of potential institutional support, evade the system through concealment and self-organize. Here, the refugee exists, in the gaze of the state, as a criminal, positioned by default to be apprehended by law enforcement.

These dynamics unveil what is at its core a racialized neoliberal control regime. Going through the official process is a gamble for potential refugees, who may find themselves registered and processed through the official resettlement program, only to have their case rejected. With a permanent unsuccessful immigration record, the only recourse for sanctuary is precarious, under the radar, and "illegal." In this sense, the "undocumented immigrant" is very much a political construction, emerging in these gaps of institutional dismantlement. Conversely, in the event of immigration being granted, newcomers are positioned as "public charges" in the racialized imaginary of the American neoliberal security state, simultaneously scapegoated for their dependence on public services while constantly in the act of being and becoming "self-sufficient."

On December 12, 2018, I learned that Reset would no longer be allowed to resettle refugees from 2019 onward.[19] "The oldest law in the United States is asylum. America was founded upon asylum-seekers fleeing oppression from the Crown," Anita notes in disbelief during a meeting where the organization is trying to formulate next steps. This was the first time since the 1990s that Reset had involuntarily foregone funding for R&P. Anita and her team of caseworkers and former refugees, including Dino, Georgiana, and Yunus, were determined to find alternative sources of housing that would adhere to the strict criteria required for eligible resettlement accommodation.[20]

DECONSTRUCTING ELLIS ISLAND ROMANTICISM

"Give me your tired, your poor, / Your huddled masses yearning to breathe free." These immortal words by Emma Lazarus, inscribed on the Statue of Liberty in 1903, have captured public imaginations about US tolerance and liberalism for more than a century.[21] In January 2018, James Comey, former FBI director under the administration of Donald Trump, tweeted these words in a rebuke to xenophobic slurs made in public by the president.[22] Harking back to an alleged golden era of tolerance and inclusion, the words of Lazarus, however, obfuscate a long tradition of institutional xenophobia and anti-immigrant sentiment.

New York, home to the poem and Lady Liberty, long held as the gold standard for migrant-receiving cities, has nevertheless been a site of aggressive immigration enforcement, assessment, and control. While the timing of my presence at Reset coincided with a historically virulent clampdown on immigration and refugee services, the disaggregation of the xenophobic migration control regime through targeted institutional dismantlement in New York can be traced back a century before Trump.

Ellis Island, the port of arrival up until 1954, was a space of continual assessment from the moment of arrival. As immigrants would step off boats and make their way to the arrival halls, they were subject to surveillance, assessment, and processing. Staff would observe new arrivals as they walked up the stairs to the hall, marking "human defects" upon entry.[23] Physicians would even observe migrants carrying luggage, looking for signs of shortness of breath and cardiac problems; they would also scan the crowds for signs of goiter, skin rashes, and trachoma.[24] These logics observed newcomers as suspicious by default. The policing of their bodies through literal symbols drawn on them is an early form of algorithmic classification, sorting entrants per the demands of American capitalism and racial regimes.[25] For example, immigrants were consistently associated with "germs and contagion" from the late 1890s into the twentieth century.[26] In the last few decades of Ellis Island, however, these logics were starting to find their way past the physical port and into cities and the national imaginary through emergent anti-immigration laws and policies.

Following the 1918 flu pandemic, for example, increasingly sophisticated health surveillance tactics were used to asymmetrically target ethnic minorities in New York City.[27] The Immigration Act of 1917 instituted a literacy test and increased payments upon arrival, while instituting a ban on individuals from "Asiatic barred zones." By extension, the Emergency Quota Act of 1921 introduced further immigration caps by nationality, while the Immigration Act of 1924 reinforced quotas and barred entry to a list of characteristics that deemed certain individuals as "ineligible for citizenship." While fewer immigrants entered the United States between 1924 and 1965, the disease connotation and xenophobic narrative intensified, exacerbated by fear of foreign insurgence and communism.[28]

It wasn't until the World War II effort brought forward by President Harry Truman, emerging out of his special directive (hence known as the Truman Directive), that a slightly more favorable approach to immigration emerges—and in addition, a guarantee of a certain amount of financial aid for refugees.[29] Even then, the Truman Directive's objective was twofold: establishing the United States as a receiving country of refugees, while also preventing them from becoming a financial burden on the state (i.e., a "public charge"). The Refugee Act passed in 1980, not inconspicuously against the backdrop of the rise of neoliberalism and a reinforced emphasis on the "self-sufficiency" of refugees and immigrant populations broadly.[30]

In 1954, Ellis Island closed, and the first notable border operation beyond the border itself debuted. Operation Wetback saw approximately four million Mexicans, documented and undocumented alike, "rounded up in factories, restaurants, bars, and even private domiciles and then expelled."[31] Paradoxically, this came just a decade after the Truman Directive, and three years after the United Nations ratified the 1951 Refugee Convention, which the United States did not sign at the time.[32] It wasn't until 1965 that the United States purportedly relaxed its attitudes toward immigration, under the Hart–Celler Immigration Act. As the United States tumbled toward the neoliberal era, the state was hollowed out and self-sufficiency at all costs became the dominant paradigm. Lest the state interfered, it became possible to imagine a place for refugees and immigrants in the American imaginary. In 1980, the Refugee Act passed, but not without significant backlash (despite the reduced public spending on refugees and immigrants). In the wake of AIDS, nativism was on the rise in the country, and so for the next ten years the state was given the discretion to exercise formerly exclusionary policies. For example, the AIDS epidemic helped nurture the insistence on restricting state services, including health care, for the undocumented and undesirable. Invasive health screenings far beyond arrivals were instituted at this time as well, under the auspices of fighting tuberculosis.[33]

Nevertheless, provisions for granting funds to "public and private non-profit agencies for initial resettlement (including initial reception and placement with sponsors) of refugees in the United States" became a reality, as a standardized reception and placement program was established.[34]

The DOS assumes the role of a creditor to local R&P providers; nine VolAgs across the country are invited to apply for funding toward refugee and immigration services, with each VolAg funneling the grant to their respective RROs.[35] Under this new resettlement regime, the ability of VolAgs to provide for their clients depends entirely on two factors: the amount of funding provided by the DOS (which is subject to a history of successful resettlements in years prior, within the VolAg and its network); and the ability of RROs to resettle and house incoming refugees within their designated communities.

Over a century of xenophobic policies and antipauperism, sorting the deserving from the undeserving, gave rise to what I describe in chapter 4 as the *anti*sanctuary city. Migration control in the United States was disaggregated in concurrence with targeted institutional decay, tightening external boundaries while erecting internal ones (where there was previously a possibility for marginal reprieve). Although Ellis Island was abandoned, its anti-immigrant logics continue to shape the immigration landscape in New York today.

NEGOTIATING THE FOREIGNER IN BERLIN

"The War Is Over—Syria Needs You" reads a poster on a bus stop near Alexanderplatz, in Berlin. The poster is one of many plastered to legitimate ad spaces by the so-called Identitarian Movement (Identitäre Bewegung). While the organization has been placed under heavy surveillance by intelligence services and suspected of being unconstitutional for its supremacist activity,[36] it is by no means fringe in either the German or the European discourse on refugees and migrants. Since 9/11, European narratives undergirding racial discrimination and differentiation have been advancing along largely cultural lines, including through prevailing Islamophobic tropes, stereotypes about either the inherent "backwardness" and/or animosity of the Middle East toward the West, or linguistic preferences.[37] This flavor of racial regime is especially apparent in transient spaces such as airports, across news media outlets, and consistently in the context of refuge. The question of the nonassimilability of the European Muslim and the refugee has come to the fore at a time when Europe—and in particular Germany—is once again in need of a young, highly skilled labor force. By 2015, however, 81 percent of asylum applications in Europe were distributed as follows: 35.2 percent in Germany, 13.9 percent in Hungary, 12.4 percent in Sweden, 6.8 percent in Austria, 6.6 percent in Italy, and 5.6 percent in France.[38]

In the 1960s, "politicians and scholars alike" had "hailed the transfer of 'unemployed' workers to labour-demanding regions as precisely the intergovernmental program needed to solve Europe's economic problems."[39] Notably, since 2000 Germany has experienced a shortage of labor in the technology sector, sparking the debates that subsequently led to passage of the Migration Act of 2005.[40] When a pilot program aimed at hiring twenty thousand "highly qualified workers from abroad" received under half the target in applicants, however, German politicians wondered whether the "good" immigrants found distasteful Germany's preservation of its "pure ethnic national identity."[41] Yet, without immigration to make up for the shortfall of some fifteen million workers by 2050, due to a shrinking, aging, and inappropriately skilled population, the country would experience economic turmoil and a slimmed-down chance of ever being a competitive actor in the IT industry.[42]

Two decades on, this remains a pressing problem. Under the chancellorship of Angela Merkel, Germany attempted to situate itself as the European alternative to Silicon Valley, particularly as it sought to, in sporadic fashion, digitalize its paper-based bureaucracy and become a competing economy in the space of artificial intelligence in particular.[43] Even with €3 billion designated for the sector by 2025, as of 2018 Germany's primary obstacle—and that of Europe, more broadly—has been to fill the so-called digital skills gap.[44] In 2019, a European Commission report found that a staggering 42 percent of Europeans were unable to perform basic digital tasks.[45] This, despite an aggregate investment of €4.3 billion in "newly-launched companies" in Germany in 2017.[46] As "a magnet for founders and investors," particularly with the departure of the United Kingdom from the European Union,[47] Berlin could still succeed London as Europe's tech capital (as a function of the quantity of start-ups). To do so, it must meet the gargantuan challenge of training and recruiting a highly skilled workforce.[48] Germany's dilemma is one rooted in a long history of negotiating proximity between the migrant guest worker—the *Gastarbeiter*—and the German citizen who depends on the productivity and taxes of the former for survival. It is not a coincidence that Merkel temporarily suspended the Dublin Regulations in 2015, opening the German labor market to eight hundred thousand Syrian refugees.[49]

Revisiting the development of German immigration policies, particularly in the postwar period, it becomes clear that the country depended on racial capital—that is, the surplus value generated from a population of workers at the margins of German society—in order to grow. Of particular importance remains the unsettled question of the permanent migrant worker, who at once agitates the German working class for driving wages down, while also not being allowed social mobility lest their returns outweigh those of ethnic Germans.[50]

"MIGRATION NOT OF CONVICTION BUT OF EXPEDIENCY"

Immigration law in Germany is a relatively new phenomenon. It wasn't until the Migration Act of 2005 that Germany moved from what might be broadly summarized as "foreigners' law" to "immigration law."[51] Since 1913, "foreigners" have been "constitutionally excluded" in Germany, where citizenship was understood as an ethnocultural phenomenon in dialectical opposition to the so-called immigrant.[52] Yet, particularly in the postwar moment, foreign labor remained a "solution to two urgent and mutually reinforcing developments: the unexpected industrial boom of the so-called *Witschaftswunder* and the growing shortage of able-bodied German male workers."[53] The war had squeezed the labor supply of European young males. As a result, nearly six hundred German labor recruitment offices were established in Mediterranean countries to drive migrant workers north: "The movement that ensued was to be a migration not of families and whole populations, but of solitary workers; a migration not of permanence, but of assumed short duration; a migration not of conviction but of expediency."[54]

Two major developments in German migration contributed to some elation about the potential for foreign labor to support German growth: first, migration into West Germany in the 1950s, as some 4.5 million refugees, expellees, and emigrants with German heritage, known as "ethnic German resettlers,"[55] fled communism; and second, the *Gastarbeiter.* Guest workers began arriving upon the successful ratification of a treaty arrangement with Italy, which was subsequently supplemented between 1955 and 1968 with similar arrangements with Spain, Greece, Turkey, Morocco, Portugal, Tunisia, and the former Yugoslavia. The operative word here is *guest*, as it indicated an expectation of return, following what was essentially a temporary work permit.[56]

In 1965, the Act on Foreigners was instated, which, for the first time, omitted any distinction between "ethnic resettlers" and guest workers. Instead, it implemented a devolved approach that left it up to administrative agencies and local courts to formulate policies interpreting and applying the act to each case. The parameters for determination were left woefully vague, stating that "a residence permit should be granted if the 'presence of the foreigner does not compromise the interests of the Federal Republic of Germany.'" It also referenced the Convention Relating to the Status of Refugees of 1951, granting that "foreigners who were granted asylum had a legal right to a residence permit." In 1969, the Law on EEC Residence was passed, which aimed to implement the law on European Economic Community workers' freedom of movement. By the following year, 25 percent of West German "foreigners" were from EEC member states.[57]

At the conclusion of the migrant programs in 1973, however, guest workers had started the process of permanent immigration, leading the German government to start "a formal policy of 'integration.'" Up until 1973, Europe's economy had been animated by "the planned migration of millions of 'temporary' workers" from the Mediterranean and throughout this time,[58] and Germany "generally considered itself a 'labour recruiting' as opposed to an 'immigration' country."[59] Nevertheless, despite awareness of the permanent residence of migrants, the country remained in denial about its changing state. Beginning in 1974, incremental shifts in policy to encourage return or constrain successful immigration were made.[60] For instance, work permits were denied to any children of foreign workers arriving after December 1, 1974. Rist recalls that "many young people from immigrant families [had] now completed their schooling" but would be ineligible for a permit, should they decide to join a parent in West Germany.[61] Children of working age were therefore faced with the prospect of being unemployed, resorting to informal labor, or facing deportation (also a consequence of taking up informal labor): "It is clear that the social policies specific to the guest-worker situation are in conflict with one another; German authorities stress reuniting families, but the youth who come are denied labour permits.'"[62]

The Return Assistance Act of 1983 was the most aggressive move to expel guest workers. Those who had arrived from countries outside the EEC would receive 10,500 DM for leaving, adding further incentive by reducing the guaranteed amount by 1,500 DM per month overstayed.[63] In other words, guest workers, on whom the German economy depended, were penalized for attempting to build lives alongside their livelihood.[64]

WILLKOMMENSKULTUR AS RACIALIZATION

The guest workers of the second half of the twentieth century were emblematic of an ever-present dilemma for both policymakers and unions in Germany. On the one hand, "the presence of highly visible immigrant groups, inferior socially and economically to the national populations, [created] a challenge to the liberal ethos of national governments committed to democracy." Rist compares this to the dilemma faced by the United States, their purported regard for democracy, and the systematic segregation and disenfranchisement of Black and Indigenous communities in particular. This marginalization must continue for migrants to remain profitable—but it also reveals the hypocrisy of the values the two nations allegedly operate under. On the other hand, should Germany choose to intentionally "create opportunity and effect social mobility for these groups, the utility of the immigrants [would be] vastly reduced and there [would be a] simultaneous rise in resentment on the part of the national working class." For unions, the dilemma is between opting to advance the cause of the German working class, which would situate them antagonistically toward the import of migrant labor, or acknowledging that "guest workers enhance the economic well-being of these same union workers."[65] The paradox for both policymakers and unions is that the guest workers are a racialized stratum of the working class.

In 1990, Germany enacted a new Act on Foreigners that doubled down on the restrictive components of the original from 1965. It was determined that "Germany's capacity to take in immigrants was not unlimited and preference had to be given to immigrants of German heritage, foreigners fleeing political persecution, and EU citizens taking advantage of freedom of movement."[66] This came despite increased arrivals of asylum-seekers between late 1980 and 1992, following the war in the former Yugoslavia.[67] Under the auspices of guaranteeing the immigration of those already "legally" residing in Germany, high court decisions that upheld Germany as a destination for foreigners (but not as an immigration country) were codified, while expulsion rules became more severe, in particular for individuals arriving from outside the EU.

At the time, Germany was able to keep its approval rate for asylum applications at 4.3 percent, following the enactment of the Asylum Compromise of 1992, which stipulated that "applicants that arrived at Germany's borders from another EU [or] neighbouring country did not have a right to asylum and could be refused entry."[68] The Federal Constitutional Court had also recently reversed decisions made by state governments in Berlin, Hamburg, and Schleswig-Holstein to grant foreigners voting rights in local elections. The ruling stated that the lower courts violated "popular sovereignty," which was understood to be only "represented by the people of the state (*Staatvolk*), prohibiting aliens to participate in elections."[69]

These moves were made not in small part due to the increasing anti-immigrant sentiment across Germany. The virulent rise of neo-Nazis led to arson attacks and pogroms against asylum-seekers, refugees, and Turkish families in the early 1990s, particularly across the eastern regions.[70] Debates facilitated by far-right parties (including the Deutsche Volksunion, Nationaldemokratische Partei, and Die Republikaner) on "'bogus' and 'fraudulent' asylum applicants . . . whose main intention was to invade and exploit the German welfare state" won neo-Nazi sympathetic politicians parliamentary seats, even in supposedly liberal Berlin.[71]

The German immigration system has historically tended to veer conservatively, to say the least, at times even legitimizing and enabling violent anti-immigrant sentiments. That is, until the need for labor imports outweighed the need for ethnonational integrity. As mentioned above, the Migration Act of 2005, which designated Germany as an "immigration country" for the first time, materialized largely against the backdrop of a shortage in skilled labor in the IT sector.[72] Following the failure of the skilled labor program (which sought to recruit twenty thousand individuals but received only eight thousand applications), the act came into force on January 1, 2005, amending the Nationality Act[73] and introducing the Residence Act,[74] which together provided a pathway to "long-term permanent residency for migrants, in particular for skilled workers." "Integration" policies were also introduced under the principle of "support and challenge," mandating obligatory classes for learning the German language and culture. Over the next decade, Germany experienced its largest population growth since 1992—out of necessity and as a result of capital input needs, as opposed to a positive national inclination toward immigrants.[75]

It should therefore come as no great shock that Merkel's response to the 2015 refugee crisis was so controversially received.[76] Virulent opposition to the arrival of refugees was particularly heightened following the New Year's Eve events of 2015–16,[77] as centrist and far-right politicians and agitators entered an implicit consensus, drawing a false equivalence between terrorism and "a failure to manage immigration" particularly from "Islamic cultures." Aleksandra Lewicki notes that as recently as 2016:

> 38.15 per cent of the people living in the former West and 53.82 per cent of the population in the former East would "prohibit Muslim immigration to Germany," while 50.3 per cent in the East and 49.92 per cent in the West claimed to "feel like a stranger in their own country due to the high Muslim presence."[78]

Sentiments such as these did not emerge overnight; they were structurally anchored to a long history of racialized immigration policies. Policies effectively treated foreign migrant labor as racial capital to be invoked economically in times of labor shortage,[79] and politically when the ethnocultural integrity of Germany was threatened.[80] This was made consistently clear as the "spectacle" of the refugee crisis was articulated in racialized terms in German public discourse,[81] and is testament to the continued "coloniality of migration."[82] Right-wing populist parties, in particular Alternative für Deutschland (AfD), rose in the political hierarchy in tandem with the anti-Muslim mobilizations of the far-right Pegida movement in late 2015: "Racist violence quadrupled to alarming peaks in 2015 and 2016. Authorities registered 970 assaults targeting refugee facilities and 2,400 attacks aimed at individual refugees in 2016."[83] In 2017, AfD arose to become a third-largest party, winning 12.7 percent of the total vote,[84] thus reconciling German democracy with its hostile immigration environment through parliamentary representation.

Finally, Germany's two most recent pieces of legislation show that even the Migration Act is subject to proclivities toward temporary residence. With the Integration Act of 2016, "Support and Challenge" was reemphasized, while adding a further infantilizing incentive structure: "refugees who show the potential to integrate" would have a good chance of permanent residence, with "easier and faster access to integration classes and employment opportunities, while refugees who refuse to cooperate face a reduction in benefits."[85] Meanwhile, the 2019 Asylum and Immigration Policy introduced the Geordnete-Rückkehr-Gesetz (Orderly Return Law), which vastly expanded police and immigration authority powers to facilitate a more concerted effort to deport "failed" asylum-seekers.[86] The Interior Ministry's objective has been clear: to improve the success rate of deportations, as roughly half of the 188,000 planned since 2015 failed or were not carried out.[87]

Throughout this chapter, we've sketched how essential migrant workers are to the survival of Germany's economy, yet the "mixed blessing" of migrant labor also exposes the country's continued struggle with structural racism.[88] Today, hardly any distinction is made between "refugees" and "labor migrants" in Berlin,[89] as the former's function is largely regarded as economic. The Identitarian Movement's posters appear as an abhorrent contrast to the urban artwork across Berlin, but they are mutually constitutive with the many civil society organizations that participate in the culture of welcome (*Willkommenskultur*). This draws some semblance to David Harvey's "accumulation by dispossession," with the caveat that immigration status is understood as a prerequisite to capital.[90] As I argue in chapter 5, *Willkommenskultur* is the vehicle through which *refugeeness*—an abstraction and essentialism of refugee subjectivities—becomes a means by which German generosity is communicated, while it remains undeniably hostile terrain. Through refugeeness, and *digital refugeeness* in particular, the veil of liberalism performs as a silent consensus appropriate for the times. By abstracting the refugee from their refugeeness, one can claim to improve their tragic conditions through measures that do not threaten either the "native" working class or the nation's need for cheap labor, while attracting venture capital and funding in the process.

4

THE DIGITAL ANTISANCTUARY OF NEW YORK

"I don't know why they call this a city of immigrants," says Maryam, nodding to the files of clients sitting on her desk—clients dreaming of refuge in New York City.[1] Against the backdrop of Trump's tightening immigration climate, visible not just by virtue of his emphatic border wall fantasies and the increasing deportation raids by ICE, but also by how the State Department has continued to defund and sabotage voluntary agencies (VolAgs)[2] in charge of providing resettlement services. Like the one Maryam works at in Brooklyn, these essential private organizations are the bedrock of resettlement in the United States. For migrant populations in New York, they are often the last hope for institutional sanctuary from persecution, poverty, and homelessness.

This sanctuary city, however, is shrinking—and not in small part due to these obstacles. By late November 2018, one of the city's largest resettlement agencies had been defunded to the point of obsolescence for resettlement and placement (R&P) services. It remained in tight competition with other VolAgs in New York, struggling to demonstrate its successful resettlement of asylum-seekers and refugees in the year past (diminished in numbers by Trump's so-called Muslim ban, which led to fewer documented individuals declaring themselves in New York). In other words, R&P under Trump was designed to fail. Trump's hostile immigration architecture is not solely based on policy, but also inherited from Silicon Valley. As the sanctuary city shrinks, displaced individuals—documented and undocumented alike—are deterred from seeking institutional support. As a result, marginal smart city services are relied on to maintain New York's symbolic status as a city of immigrants. Yet, many of these digital infrastructures are "in complete contradiction of the sanctuary city."[3]

Celebrated digital pathways to sanctuary in the city—such as the public Wi-Fi project known as LinkNYC, IDNYC's municipal ID system, and the affordable housing initiative Housing Connect—are just a few of the tools weaving precarity, fear, and surveillance into the fabric of urban migrant life. Marginalized communities on the move are pulled into the digital periphery in New York through their categorization as needy (requiring technical solutions to their predicament), criminal, and elusive (justifying their identification, monitoring, and surveillance), leading to their containment through limiting mobility and access via information panics or direct interception, detention, and deportation. As the city has become layered with an information control architecture that mediates access to crucial services, while reinforcing the disaggregated panopticon of a hostile immigration regime in the urban milieu, newcomers and mature immigrant communities alike have also been increasingly engaged in refusing datafication through urban concealment.

In the sections that follow, I sketch out New York as a site of mundane border enforcement beyond the border. Short- and long-term ethnographic encounters with public defenders, caseworkers, activists, and self-identifying migrants of varying status have shaped my framing of the city as a digital antisanctuary. These encounters elucidate how the emergence of digital forms of urban migration control—observable through experiences of technology-driven fear and precarity in vulnerable migrant populations—demarcate the digital periphery. The deference to technological solutions in realms crucial to everyday life—such as access to information, identification, and housing—permit technology giants to play a subtle yet increasingly active role in the control of undesirable migrant bodies. In exchange, cities such as New York continue posturing as sanctuaries, while facilitating the rapid and lucrative weaving of Silicon Valley into the fabric of urban governance.

THE ANTISANCTUARY CITY

I arrived in New York City in a moment of particular turmoil for immigrant communities at large; US Secretary of Homeland Security Kirstjen Nielsen had signed off on "option 3" in April 2018—a deterrent strategy set to curb migration by putting in motion the indiscriminate prosecution of "every adult who crossed the border illegally, including those who came with children."[4] While DHS has ramped up its family separations, it also emerged that families were unable to reunite due to technical shortcomings that prevented parents from being digitally "tagged" with their separated children. [5] Trump's border now operating with the ability to digitally erase migrants altogether, "transgresses" families and fortifies the artificially imposed divides between them.[6] The digital augmentation of the US border, however, did not emerge overnight, nor was it limited to the purview of border enforcement agencies such as ICE and US Customs and Border Protection.

As Mike Davis teaches us in *No One Is Illegal*, "what truly demarcates the United States is not so much the scale or frequency of state repression, but rather the extraordinary centrality of institutionalized private violence in the reproduction of the racial and social order."[7] The urban, and the smart urban in particular,[8] has played an instrumental role in operationalizing this order through technology. Positing a critique against the celebrated post-racial turn in São Paulo, Jaime Amparo Alves frames urban practices of segregation, economic exclusion, mass incarceration, and police violence against predominantly Black youth as premier features of the "anti-black city."[9] Similarly, I assume a critical position against celebratory discourses around New York as a "city of immigrants," encountering it instead as an antisanctuary city. Practices underscoring the antisanctuary demonstrate what remains a fraught battleground rife with struggles for immigrant justice and refuge from persecution, against the backdrop of an expansive and often hidden consensus among private technology actors, the city, and federal agencies in the urban reinforcement of anti-immigrant marginalization, violence, and precarity.

I meet David in Crown Heights, Brooklyn, one September afternoon; David's organization, the Brooklyn Defender Services (BDS), operates out of the neighborhood.[10] Crown Heights—which had been previously occupied by mostly immigrants of Jewish, Irish, and Italian origins[11] and considered 70 percent white by 1960—had seen considerable white and capital flight as immigrants from the West Indies and the US South began to settle there (as well as East Flatbush, Brooklyn; the North Bronx; and Laurelton, Queens).[12] By 1970, the demographic makeup of Crown Heights had changed significantly, and it is now considered 70 percent Black. Today, Crown Heights remains a destination for newcomers; the neighborhood houses enough immigrants to warrant the attention of ICE and (Customs and Border Protection), who were "filmed knocking on apartment doors" as recently as March 2020.[13]

David works with families subject to child neglect charges; particularly damning for the undocumented, these charges are often inextricable from the families' particular immigration predicament. As undocumented families, David's clients experience severe obstacles when accessing social support systems and services.[14] Similarly, the lack of institutional flexibility in considering language and cultural barriers means that Child Protective Services adopt an agnostic attitude toward the particularities of the case.[15] The charge itself is typically a result of irregular working hours transpiring from poverty and the lack of access to formal work. To exacerbate these challenges, David tells me that families subject to child neglect charges are at high risk of being flagged by ICE, thus risking their detention and deportation.[16]

He describes the profoundly unknowable nature of the immigration environment in New York. For example, he is concerned about his clients showing up to family court, only to be detained by ICE: "New York City has sanctuary status, but this is constantly under question. There remains a worry that even if a trial is merely registered in the court's calendar, that there is some way that ICE might use that data too." Previously, David was able to reassure his clients about risks associated with ICE, which he says used to be practically nonexistent once they reached out to Defender Services for help. Increasingly, however, there is an emerging tension between information solicitation and surveillance. As a lawyer, David is severely constrained in his ability to serve clients without access to data. Finding himself unable to ask for compromising information, he is forced to advise clients to plead to certain charges to avoid the creation of risky and potentially identifiable data trails. "Family court seems to be the typical entry point," he says, referring to data that ICE might obtain. "The bar is very low for undocumented immigrants. It is usually for benign things such as a child in school mentioning that they were depressed," for which families are held accountable by default. As Ana Muñiz's work shows, immigration enforcement in the United States is undergirded by data systems with an ever-expansive capture of possible "criminal" or "dangerous" immigrants, using flawed and unreliable labeling practices.[17] That family court might be flagged under similar labels is hardly surprising. In response, David's clients, along with most newcomers I spoke with in the city, preferred to stay invisible within diaspora networks.

It has long been established that refugees go to cities to be closer to potential familial networks and live anonymously with their support.[18] "There's a place for everyone in New York, but you have to make that place on your own," says Majeed,[19] a Chadian asylum-seeker in his mid-thirties who I had met during a support group at Human Rights First, lamenting the jarring and disorienting experience of integrating and belonging in the city with no one to reach out to. He did not have connections to the diaspora through family or friends of friends, unlike other attendees such as Abbas and Talal, who had been provided with both housing and work in Flatbush.[20] Majeed, however, had been the subject of a great deal of serendipity, as his first encounter with a Chadian taxi driver provided him with a direct link to the diaspora. He was driven to a house in New Jersey, where he was met by his host, an American Chadian man. "Us from Chad need to be brothers and help each other," he recalls being told, before being put up in the house with the man's family. Majeed recently moved into a larger shared apartment with other people, allowing them to pool resources; his eventual job as a grocer, referred to him by his former host, had provided some financial security, against all odds.

Majeed's story, however, is by no means straightforward. While newcomers rely on anchor communities they may know in advance or seek out upon arrival, survival provisions are neither guaranteed nor necessarily consistent. The resignation of newcomers to often chaotic but concealed diaspora spaces, hidden from the gaze of the state, means there is little if any institutional support, which is particularly challenging for families. Declaring your intention to apply for asylum, however, risks putting you on ICE's radar and a path to detention and potentially deportation. As David explains, immigrant invisibility is disrupted when you come to court. He recalls how going through the formal immigration system, in one instance, had led a client to self-deport: "They had come to America because their daughter had no future in Guatemala. The paperwork was grueling. They chose to return."[21]

Given the hostility of this environment, I wonder how David can continue his work. He pauses and looks somewhat resigned before replying, "By building trusted relationships—not with forms, but with humans," recalling that even in the human case, "Documenting truth is potentially risky to the immigrant." What is at stake in New York is not simply a new immigration enforcement paradigm, but a weaponized information environment, requiring those who are subject to it—immigrant populations and service providers alike—to iterate emergent tactics of urban concealment. "DACA [Deferred Action for Childhood Arrivals] recipients have been consistently screwed over; they registered and gave details about how they arrived in the US under the impression that they were guaranteed a path to citizenship," he says, regretfully.[22] The Obama-era program made individuals brought in to the United States as children eligible for legal status (pending high school graduation or honorable military discharge and a background check), but became subject to dismantling by the Trump administration. While federal courts ordered the administration to continue processing cases, the order applied only to existing cases rather than new ones.[23] In the wake of the Trump regime, this data also put them in danger of being deported.[24] This is no different for asylum-seekers, whose "legitimate asylum-claims are being upended, and data volunteered on that front is also used against the applicant."[25] Despite greater socioeconomic precarity, undocumented immigrants must resort to informal work under the shelter of the "invisible diaspora."

DIGITAL AFTERLIVES OF THE BORDER WALL

Tech offerings are included as crucial elements in the outreach work of the Mayor's Office for Immigrant Affairs (MOIA), including its provision of translation services, immigration services, connectivity, and identification. Infrastructural technologies such as public Wi-Fi, through the LinkNYC kiosks, municipal ID systems (IDNYC), and affordable housing platforms (Housing Connect) have been strongly featured elements of MOIA's strategy. Against the backdrop of ICE crackdowns on undocumented communities, however, my informants warned me that public Wi-Fi was both leading to information panics among vulnerable immigrant communities and potentially correlated to detentions and deportations. Even as the city officially refuses to share immigration status–related data with the federal government, it is unclear whether the same can be said for the technology contractors working with institutions including but not limited to the Mayor's Office for Immigrant Affairs, the Department of Housing Preservation and Development (HPD), and the New York Police Department (NYPD). New York as a smart city reveals two logics of antisanctuary: information panics and urban instrumentalization.

"You know, Church Avenue has the largest concentration of undocumented immigrants?" caseworker Anita asks me. "It's also where we have the one LinkNYC kiosk in the area! We were so excited at first; people can charge their phones, get Wi-Fi, and come join us . . . we have our Cooper Union [program] ad on the kiosk."[26] Since I first arrived, the city, and MOIA in particular, had been subject to mounting critique over the NYPD and ICE officers' deployment of high-tech surveillance practices, including but not limited to license plate readers, a centralized app for monitoring security cameras across the city (part of the Domain Awareness System), and other predictive analytics tools used to determine the likelihood and presence of undocumented immigrants. Activists were already calling on MOIA to do more to stifle the capabilities of ICE. The LinkNYC kiosks Anita refers to are the product of the city's contract with the Sidewalk Labs daughter company Intersection, which is responsible for the delivery of what purports to be a free public Wi-Fi project.[27]

Payphones across New York had been replaced by the nodes, delivered in partnership with the Mayor's Office and Intersection. Each kiosk is equipped with public Wi-Fi, the ability to place phone calls and charge devices, and a tablet through which city services, maps, and directions can be accessed. New York is littered with the towering screens, which give recommendations about what to see in the city (or at the time of writing, list the names of victims of police brutality). The kiosks also display advertisements and fun facts tailored to the particular area. Intersection has deep partnerships with MOIA as well as the NYC Commission on Human Rights, giving access to immigrant services (e.g., NYC311 and translation services) on the kiosks. It has also been known to provide ad space for local immigrant rights organizations and services.

Contrary to how the initiative has been advertised, however, each unit also has three hidden cameras (these are fairly obscure unless you look closely behind the black panel) and a Bluetooth sensor. This gives rise to several concerns and points of inquiry. First and foremost, Intersection currently holds more than 8.6 terabytes of user data distributed over six million unique users in New York[28]—this includes extensive device data, not limited to language, browser type, time zone settings, and so forth. This is population data that advertisers can use to target particular demographics and communities in New York. This is also data that is potentially detrimental to affected communities of immigrants, particularly in the wrong hands. In 2016, Charles Meyers exposed code from the public GitHub repository of LinkNYC,[29] which indicated that the kiosks were capable of collecting and organizing users' geocoordinates, browser type, operating system, device type, device identifier, and full URL clickstreams (including date and time). Intersection's own privacy policy shows that the following data is collected: MAC address, IP address, browser type and version, time zone settings, browser-plugin types and version, operating system, device type, and device identifier.[30]

At the Intersection offices, the concerns and resistance expressed by communities remained a puzzle with no straightforward technical fix. Isabel, one of the project leads for LinkNYC at Intersection, describes how neighborhoods "where people don't normally get services" tend to harbor suspicions around the costs of the kiosk: "They're like, what does this cost mean? Nothing is free."[31] Intersection had toyed around with releasing a downloadable app that could provide more info on the kiosks and centralize city services, but quickly realized that privacy concerns remained salient in the city and people were therefore less likely to download it. When I ask about other ways to further inform residents about the "fine print" of the kiosks, I am told that "we do have an FAQ on our website, but it's only in English."[32]

I am puzzled following subsequent revelations by Isabel that Intersection proudly hosts a "how to connect" site translated to six different languages. It is telling that the barriers to entry to LinkNYC are significantly lower than the barriers to exit. Isabel (external comms), Zeynep (back-end developer), and Mike (data analytics) were mid-level employees tasked with deploying the kiosks:[33]

> We worked with local nonprofits, to let them know they could use the space for putting their content. Making sure individuals in the communities get platforms to talk about issues that matter to communities. Reactions have been quite excited. Not only are we providing a new technology and offering a free product, but we're also on the streets with LinkNYC, engaging with campaigns such as DoSomething.org, JustFix, the gun violence campaigns, and Pride.[34]

Isabel's team of technologists appeared genuinely convinced that they were working on behalf of communities. This made it all the more concerning that they weren't in any meaningful way plugged into wider critical discussions about data-sharing and protection practices within the organization.

The Intersection workers I spoke to were adamant that they let people know that their devices are not allowed to have facial recognition; that all footage clears after sixty days; that the only personally identifiable information collected is technical, consisting of unverified email addresses, device type, phone language, and a unique hash key to be able to automatically reconnect devices to their network. Yet, these assurances hardly allay major concerns around LinkNYC, which were threaded through most of the perceptions of organizations and migrants I spoke to. These included uncertainties around (1) data being collected through kiosks—across the board, communities were concerned about audio and video content being collected via the hidden cameras, in particular, and stored for any time, let alone sixty days; (2) how that information was processed; and (3) law enforcement having access to this data.

Anita had only learned about these concerns after having been in consultation with Intersection and agreeing to advertise Reset services on them: "We have no time to take a break and look at these things; we're in a position where we're having to behave like poor people and accept bad deals with high interest." Anita holds up the key fob to the office doors, situated between the reception and the cubicles where caseworkers work: "These key fobs are about keeping information inside; there are vulnerable clients here, with health and other issues we wouldn't want getting out." The doors separating these spaces were designed to automatically lock every time they closed, to keep information contained for the duration of time clients spent at the Reset offices. On Church Avenue, however, clients would be compromised once more. As I would come to find out, data collected through the kiosks like the one outside the Reset offices, along with census data, aided in Intersection's targeting, linguistically and culturally, of communities of interest.

Alhasan, an organizer with the Darfur People's Association of New York, asked me, "How can I trust this thing when it doesn't prompt me before just connecting me?" He explained that the community tends to "trust what's close."[35] That even he, as a transit superintendent with the Metropolitan Transport Authority, "would never pick up transit or public Wi-Fi; only at family and friends' homes! We don't trust the whole T&C [terms and conditions] 'if you agree, read five pages, and send a tweet' [to get online]."

Yunus from Reset is unsurprised: "So that's one thing that is crucial right now because people are afraid of being deported. And they don't know even if their case is pending. They cannot be deported [as long as their case is pending], but they are afraid of just putting their information out." For Yunus and other caseworkers intimately aware of the concerns of refugees and asylum seekers, MOIA's digital immigration services only further widen the perimeter of an urban geography of fear. These tools meet an imagined need, which is often incongruent with the realities of those they claim to serve. Precarious communities on the move don't want your futuristic high-tech fantasy ad board; information exchange on their terms, through word of mouth, plays a much more important role.[36]

While public consultations in the design phase of the Wi-Fi kiosks involved "the community" (unclear which) in the iteration of the device, initial designs never included the surveillance capabilities that were subsequently added. Several activist organizations called for the kiosks to be dismantled or redesigned, but both the city and Intersection have been reluctant. In mid-November 2018, I met with RethinkLinkNYC at the People's Forum in Manhattan, a "movement incubator for working-class and marginalized communities" from where most of their work takes form.[37] Amelia and Melissa are organizers with Rethink. Amelia immediately lays out the stakes: "What we're battling here is a world view." As an organization, Rethink's work consists of resisting and challenging the surveillance models on which LinkNYC kiosks are built. As individuals subject to different, but intersecting forms of violence associated with surveillance—the threat of detention and deportation in the face of precarious immigration status, and the use of surveillance in domestic abuse—Amelia and Melissa were especially concerned with the emergence of the boxes in the context of the city as a sanctuary. "It's in complete contradiction with the concept of a sanctuary city," Amelia explains. While, initially, LinkNYC seemed to enhance commercial activity and community-building by allowing browser access, it disabled the feature when "encampments by homeless people started happening. [It claimed] porn was another reason. But community-building activities were not tolerated."[38]

Melissa clarifies that the rollout of the LinkNYC pilot was "not attempting to take into account how different communities move. It was a catchall game in terms of data." The kiosks were in the process of becoming a routinized part of the urban landscape of New York; Melissa underscores that "it's the process of normalization that is problematic." Since then, it has become standard practice for each device to be programmed according to the particular neighborhood in which it is located. The implications of a rollout in an immigration-dense area, as has been the case in Queens, are potentially dire. The mediation of social services through kiosks, rather than buildings, not only bars especially vulnerable individuals from the ability to speak to someone face to face or obtain access to momentary shelter while doing so, but also "damages trust."[39]

"A great show of faith would be to remove the cameras," says Melissa. As an activist, and one who organizes with and on behalf of undocumented immigrants and communities of color, in particular, she is concerned about the possibilities for resistance, in light of recent protests against Trump's immigration policies, in the future—"They know protests are happening in Manhattan; they know the uprising is coming, and they need to keep tabs on it." Amelia chimed in, describing by analogy that the dynamics of being watched affects their ability to speak truth to power, not too dissimilarly from the ability of survivors of domestic abuse to report on their abusers: "Most of us have different reasons for being viewed differently. . . . There's an aspect of not wanting to rock the boat, even if you're aware of things, out of fear that you might be targeted, or that your community might."[40] These are hard conversations; the embodied experience of surveillance-induced trauma is an inescapable reality, hidden under the allure of the city's techno-chauvinism,[41] out of sight and out of politics. This is true for asylees skeptical of data collection in light of having fled persecution, for undocumented immigrants who are fighting to maintain what little legal claim they have to the homes they have been cultivating for decades, and for survivors of other forms of violent surveillance, including domestic abuse.

"Daniel Doctoroff is someone who didn't like the city to begin with, and now he's remaking it so that his types can come here," Amelia had said at our last meeting.[42] Doctoroff belongs to the generation of classic urban "do or die" capitalists who benefited from the revolving door between politics and the corporate sector. Having first served as deputy mayor for economic development and reconstruction under Mayor Mike Bloomberg, and then CEO and president of Bloomberg LP until 2015, he founded Sidewalk Labs and commenced aggressive lobbying for a smart city future for New York. "He hates the city," she says.[43] Indeed, Doctoroff, in consort with Mayor Bill de Blasio, played a pivotal role placing Silicon Valley companies in key governmental bodies across the city, sowing the seeds that would transform New York into an unchecked test bed for technology products. Between November 2014 and my arrival in September 2018, the city announced a series of actions that would see the tech sector more firmly woven into the fabric of New York. Let's take stock of this brief history:

On November 17, 2014, Mayor de Blasio announced that a new consortium of technology, advertising, and design firms had won the city's bid to replace its expired payphones with the world's "largest and fastest free municipal Wi-Fi network."[44] By late 2015, the first LinkNYC kiosk had been installed. By 2016, LinkNYC was the only Sidewalk Labs project publicly discussed, as Doctoroff framed the kiosks as a solution to the predicament of the urban poor from the podium at the Manhattan Yale Club.[45] In January 2017, De Blasio announced that he would defy Trump's stance on sanctuary cities. Appearing live on CNN to defend his position, de Blasio laid out the risks: "The NYPD has spent decades building a relationship with communities, including immigrant communities. . . . New York City has half a million undocumented people. We want them to come forward and work with the police. . . . If they believe by talking to a police officer, they will get deported then be torn apart from their family, they're not going to work with police."[46] For de Blasio, Trump is not a threat to undocumented communities, but to the NYPD's capability to police the city at the expense of communities with precarious immigration status.

Later that year, in October 2017, De Blasio launches NYCx with the announcement that "technology is an inescapable, critical part of our lives and the future of our communities." The Mayor's Office further elaborates that "NYCx will open urban spaces as testbeds for new technologies as a core part of the program," transforming "the relationship between city government, community and the tech industry to be more collaborative and inclusive."[47] Indeed, the NYCx Advisory Board, appointed three months later, created a powerful interface between city government and the tech industry—the kind of access to experimental test sites technologists dream of. Giants including Facebook, Google, IBM, Microsoft, Verizon, and Viacom now occupied significant positions on elite city committees.[48] Finally, on September 17, 2018, shortly after my arrival in the city, de Blasio announced a partnership between MOIA and LinkNYC aimed at providing citizenship services and information to immigrant communities.[49] At this stage it is public knowledge that ICE is operating out of New York, that Silicon Valley giants are supplying ICE software that is used for detention and deportation, and that New York's smart surveillance infrastructure (built in large part by Microsoft, IBM, and Google) may be accessed by immigration enforcement officials or shared with them by the NYPD.

During my engagements with Rethink, I became concerned about the spatial coverage of the LinkNYC nodes—particularly concerning places that were considered potential hotspots for ICE deportation raids. I scraped and mapped publicly available accounts of ICE deportation raids, curated by the Immigrant Defense Project and the Center for Constitutional Rights, against an updated list of kiosks (provided on LinkNYC's website). The map that emerged in many ways serves as the spatialized representation of but a fraction of immigrant fears in the city. With coverage leaving Manhattan practically invisible and indiscriminately surveilled while forming datafied walls around poorer neighborhoods of Brooklyn and Queens, the map unveils the ways digital infrastructures, such as LinkNYC, interact with physical infrastructures, interfering in everyday human movement. Because immigration raids are understood as an instrumentalization of the everyday urban environments of New Yorkers—and therefore as a spatial phenomenon—it stands to reason that those same environments are interrogated for their violation of trust. The kiosks, under community suspicions of complicity, are regarded as extending and increasing the possibility of the raid and causing information panics.

The city has, in other words, been retrofitted with fear by design. Information panics are a direct result of the introduction of "electronic 'eyes on the street' [as] a 'solution' to the 'fear of strangers.'"[50] On the one hand, these surveillance infrastructures usually appease white, upper-middle-class citizens and, on the other hand, are weaponized by immigration enforcement. This contributes to a constant state of precarity and unrest for communities on the move, whose neighborhoods are turned into fearscapes.[51] Smart urban surveillance infrastructures remove the logical limits to the exponential potential for control and violence; what might be seen as benign localized advertising boards to some can also be used for surveillance nodes by others. These technologies are "panoptic sorts," as per Oscar Gandy, and their "work is never done. Each use generates new uses. Each application justifies another,"[52] deepening the information panic.

In early 2019, the case of Juan Rodriguez, an unhoused man accosted by the NYPD through the use of LinkNYC, would give New Yorkers a limited demonstration of how the technology might be used. I'm on the phone with Isabel yet again as she clarifies Intersection's position, claiming that Rodriguez had been arrested forty times before and "smashed forty-two screens over the past five days—so we turned on the cameras to capture him." Isabel assures me that Intersection is "working with a nonprofit to get help for this individual, as they're clearly mentally unstable," stressing that the cameras were not in fact on and were only activated once Intersection had pinned down the location of Rodriguez. "There [have] been a lot of misconceptions around this," Isabel insists. "The general public are actually pretty supportive of Link, but the people pushing back are from an activist group—very niche." Isabel insists that fringe activists had managed to get it published in the local news blog *Gothamist*, but that "it's not like it made it into the *Daily News* or anything like that."[53] In the same breath, she explains how this was an opportunity for Intersection to release a public transparency report about data handovers to law enforcement agencies. The "extended transparency report" released in the aftermath of Rodriguez shows the number of requests made by law enforcement to turn over data held by Intersection. It goes beyond subpoenas to include what agencies have requested and obtained existing footage. However, the report does not show requests made to activate cameras or monitor data activity in particular areas. What surfaced was an elaborate public relations dance between gaslighting fear of surveillance and the rationalization of Rodriguez's resistance as a "mentally ill" response.

LinkNYC is one of the many other infrastructures of surveillance that disrupt and threaten the lives of historically marginalized and targeted communities. For example, in 2021, I joined Amnesty International where, together with my friend and project colead, feminist design practitioner Sophie Dyer, we set out to extend this analysis to a larger investigation of the scope of facial recognition across New York City. With the help of some 7,500 volunteers, a group of data scientists, architects, designers, and organizers, we tagged cameras visible on Google Street View and compared them to demographic census data and stop-and-frisks among other things. The results were staggering:

- 25,500 public and private cameras were found across the city.
- The data was also cross-referenced with the NYPD's own stop-and-frisk data, showing that New Yorkers living in areas at greater risk of being stopped are also more likely to be more exposed to facial recognition technology.
- The Bronx, Brooklyn, and Queens, areas with higher proportions of people of color, had a higher concentration of facial recognition–compatible cameras.

In this sense, the correspondence between stop-and-frisk concentration and the geographical concentration of the cameras points to a persistent form of "digital frisking." This means that certain neighborhoods are more prone to exposure to facial recognition than others—and these tend to follow racial lines.

DIGITAL BORDERING IN THE CITY

"What about the MOIA's digital services for immigrant communities?" I ask Yunus.[54] He shrugs dismissively and says, "It doesn't help. It doesn't help at all. [They only help] immigrants who are [already] here and are offered citizenship. . . . Only refugee agencies are really the ones that are helping. The rest [are] just there. It's just information":

> You give me information on the computer, which I don't know how to use. You give me information about credit; I'm not worried about credit. I'm worried about paying my rent. I'm worried about getting my child to school. I'm worried about . . . you just told me that my nine-year-old cannot stay at home, well how am I going to [go to work]? Who's gonna pick up the child? Those are the challenges that are faced by immigrants in this country, which are not addressed in a way that helps.

Elaborating on his many clients who would avoid digitally mediated services at all costs, Yunus emphasizes that there's widespread "fear that their information's gonna be out, [which] is crucial right now [as] people are afraid of being deported. And they don't know even if their case is pending." In most cases, there isn't ground for deportation, but Yunus's clients tend to prefer word of mouth for this reason: "And again, there is a myth out there about accessing benefits. When you are an asylee or a refugee. Some nationalities don't want that because they feel like they will be considered as a public charge and will be deported, perhaps. Therefore, they are afraid of accessing those." Many of the newcomers I encountered in New York reiterated this fear. Attempts at upholding New York's mythological status as a city of immigrants remained firmly rooted in a techno-solutionist orientation to interventions, which did not in any way factor in community experiences of said information environment. The city's municipal smart ID program is a prime example.

The city's rollout of an IDNYC identity card for New Yorkers was initially treated as a welcomed intervention. Launched in 2015, with an enrollment of just over 1.3 million residents by late 2020, the "government-issued photo identification card" is an attempt to improve access to city services, irrespective of status.[55] As per the IDNYC website, the card is for everyone, "including the most vulnerable communities—homeless, youth, the elderly, undocumented immigrants, formerly incarcerated people, and others who may have difficulty obtaining other government-issued ID."[56] By serving as an ID that can be used to interact with the NYPD—in theory, helping migrants avoid being taken into custody not having ID—the card allegedly protects populations who are often subject to selective, biased, and persistent policing. While the card is available to everyone, it is considered especially helpful for refugees and undocumented immigrants, as a means for protecting their immigration status from being exposed. Reset sends all their clients to get an IDNYC from MOIA.[57] Requiring only a phone bill and a phone number to register, the card does not reveal the holder's address yet is supposed to be considered a legitimate form of identification by the NYPD.

Grisha, a senior activist with the nonprofit RUSA LGBTQ+—an organization that helps LGBTQ+-identifying refugees and immigrants from former Soviet nations—stresses that IDNYC hardly achieves the mainstreaming it purports to, as its recognition is still very much in question:

> So, it's not recognized by some bars, it's not . . . you cannot get credit or loan whatever, because it's only very limited—it's not Chase. It's not Citi[bank]. It's not Digibank. It's like some small banks nobody knows about. But you can go to MoMA [Museum of Modern Art], you can go to Whitney [Museum of American Art], you can go blah, blah, blah. That's good, beneficial, sure. But when people have nothing to eat, they don't care about MoMA.[58]

Yunus is equally perplexed by the city's apparent belief that immigrant communities require what both he and Grisha perceive as trivial features:

> Really? What can I do with the New York City ID? Do you know what amenity it gives you? To go to the museum. And to go to the aquarium.

For free, right? You can go to any museum to any park—it's free. But you can't open an account with it. Immigrants don't care about IDs if they do not protect them. . . . Talk to me about affordable housing![59]

Despite producing the card upon request, cardholders have been profiled by ICE and in some cases taken into custody. Reviews on IDNYC's Facebook page from October 2018 document this: "Very hesitant to use, especially after reading how much trouble it caused this undocumented hard-working immigrant delivering pizza to a local military base."[60] The post is referring to the incident involving Pablo Villavicencio Calderon, a thirty-five-year-old undocumented immigrant who was making a pizza delivery to the US Army base in Fort Hamilton, Brooklyn, in 2018. Ordinarily, Calderon would use IDNYC without a problem, but he was prompted to present a driver's license. In the absence of his license, security ran his background and found an open deportation order from 2010, at which point he was detained and later handed over to ICE.[61] Calderon now faces deportation. On another occasion, a couple attempting to visit their grandson—a US Army sergeant—were detained on July 4, 2018.[62] These are just a couple of examples demonstrating the consequences of the visibility produced through IDNYC.

This could get worse, as the city, along with MOIA, spearheaded an effort in 2019 to include financial services in the ID. Rather than extending access to mainstream banking as requested by immigrant advocates, the card is set to introduce a parallel track for financial inclusion, which has been critiqued for introducing even further risks to communities already at great risk of identification, detention, and subsequent deportation. The proposed upgrade to the card would see it equipped with an RFID, enabling it to act as a prepaid debit card of sorts.[63] This would significantly compromise undocumented immigrants in particular, as the city would be obligated to retain data (generating a centralized database on individuals or aliases attached to particular financial transactions), which may or may not be accessed by law enforcement whose relationship with federal actors such as ICE remains murky.

Furthermore, adversarial actors such as the ICE are technically capable of picking up information from RFID-enabled cards at a distance;[64] moreover, if the card communicates its description as an IDNYC card, this risks raising suspicion around the status of the cardholder.

Yet, as the city planned to add RFID and financial capabilities to the card, leading to greater data centralization, a coalition of immigrant and privacy rights organizations called for the change to be canceled. In September 2019, Councilman Carlos Menchaca (D-Brooklyn) submitted a bill that would prohibit the city from enabling RFID in the IDNYC card.[65] Meanwhile, MOIA commissioner Bitta Mostofi remained adamant that financial inclusion was a vital objective of the card, though it was at odds with the sanctuary city's duty to prevent any centralized storage of data on undocumented immigrants.[66] A letter, dated September 12, 2019, drafted and signed by at least forty-six community, labor, immigrant, civil rights, legal services, and economic justice organizations in New York, expressed its unanimous opposition to the inclusion of RFID technology in IDNYC:

> If implemented, the proposed changes to IDNYC would facilitate unprecedented, wide-scale data collection about New Yorkers' travel, spending, and other activities. Indeed, administration officials have spoken publicly about their express interest in generating "big data" and revenue through IDNYC cards equipped with smart chips. Even if well-intended, connecting this kind of technology and data to vulnerable New Yorkers' identification cards would expose people to serious risks—including dangerous experimentation or misuse by current or future administrations and private vendors—that far outweigh any potential benefits. These risks are particularly heightened given the Trump administration's escalating attacks on immigrant communities.[67]

It stands to reason that advocacy groups would voice significant concerns about any attempt to further enable the ability of authority to distinguish and identify undocumented immigrants using the card. It also calls into question the recent hype at the intersection of digital and biometric ID systems, and refugee justice broadly. Who are they for? In the case of IDNYC and similar interventions, it is clear they were designed with law enforcement in mind, not immigrants.

Other more recent high-tech offerings track New Yorkers in general, specifically their movements across the city subway system. One Metro New York (ONMY) was rolled out as a pilot in May 2019. The system was provided by San Diego–based Cubic Transportation Systems,[68] a subsidiary of defense contractor Cubic Corporation, as part of a $553.8 million contract with the Metropolitan Transport Authority.[69] OMNY currently works through NFC-enabled devices and contactless payment cards, in a bid to completely replace the current MetroCard system—a move that is bound to be met with controversy in the wake of ongoing disputes with the Mayor's Office on providing RFID-enabled IDNYC cards.

In the wake of OMNY, civil society has sounded the alarm. In particular, the Surveillance Technology Oversight Project[70] has expressed concerns that the payment system's privacy policy[71] is too ambiguous. Moreover, it is only available online, which is problematic since it governs usage in daily life that does not require a user to go online in order to consent to data-sharing. There are currently little to no assurances about the length of time indiscriminately collected data on rider travel history and usage can be retained, which runs the risk of indefinite data storage.

As it currently stands, the policy does not prevent rider data from being shared with other government agencies. A report by *The Gothamist* also found that the platforms on which devices and cards are tapped contain cameras,[72] a fact that is not at present documented and acknowledged elsewhere. (A spokesperson from Cubic claimed the cameras were only in place for QR codes and are currently not used for any biometrics.) OMNY has furthermore positioned itself as an incontestable feature of New York, pairing the act of "vandalizing OMNY readers" with unattended packages and other suspicious, conventionally terrorism-attributed behavior.[73] This comes as OMNY readers were smashed during the "J31" day of action against the NYPD on January 31, 2020.[74] While location and payment tracking on the New York City Subway system is not a fundamentally new phenomenon (this can already be done in a limited capacity with MetroCard), OMNY system increases the number of data points and the amount of data collected. As long as there remains an unclear policy on the limits of data-sharing between OMNY and government agencies, the system risks further enabling the technological anti-immigration architecture of ICE.

THE MYTH OF DIGITAL ACCESS

> As if they're helping. Can you really show me what help [they] give? The only help they give . . . in New York . . . they offer health insurance—and nobody will tell you. There's no day care, there's no after school, there is no place to live, rent, or nothing. Some people get taken advantage of by the owners of the house because they are paying a lot of money on this shitty place to stay. . . . And they make funny deals with like ten to sixteen people living in a small space upstairs. Check out the fire that happened a couple of years ago . . . in the Bronx or Harlem, two families lost. . . . Like what, eleven people died in that fire? Because most Africans—and most immigrants—we cook by way of frying, so they don't like smoke detectors. People from the Middle East don't like them either. . . . So sixteen people died in the fire because they were crammed in one apartment. Because there's no space.

Yunus is fed up. While immigrants in New York have been facing death owing to a lack of space, the Mayor's Office had been ostensibly innovating its affordable housing project. Although commendable in theory, the project but has turned out to be superfluous and obsolete in practice. "They call it Housing Connect," one of the Mayor's Office's flagship digital services aimed at housing immigrants and the poor. However, no Reset client—that is to say, no newcomer, refugee, or asylee—had been successful with Housing Connect, an observation that remained consistent across the other immigrants' rights organizations I spoke with.

During my time with Reset, Yunus regularly reminded me, with an extreme sense of urgency, that Housing Connect—should you be lucky enough to get accepted—could only provide cheaper rent bands due to subsidies received from the city. However, in the few cases where clients had been granted a tentative spot through the service, contractors had informally offered a part of the subsidy as a lump sum (approximately $10,000). "For a person in need, [that] will be very attractive, even if they will never be able to afford a place in New York with that money," notes Yunus, because "landlords prefer renting apartments at above market value, as richer New Yorkers are willing and able to pay."[75] There is, in other words, no technical fix to segregation and predatory housing practices in New York. Housing Connect layers existing systemic inequities with a reinforcing veil; under the auspices of social justice, it permits digitally augmented forms of segregation.

Housing Connect, New York's online affordable housing portal, was launched in 2012. The system is held as an in-house database at the HPD, though all the applications go through a third-party marketing agent. The housing developer receives the input information log and must go through it, selecting candidates in accordance with the marketing guidelines. Upon notification of selection, applicants are interviewed, after which there might be a follow-up interview before housing is, in theory, finalized. Developers were incentivized to create affordable housing units through a subsidy. These units were then entered into a lottery in coordination with HPD to allocate housing for low-income New Yorkers.[76] In 2012, a test website was launched to "bring the agency into the 21st century," and to improve access to the lottery. When the system launched in twelve languages, it saw a surge in the number of applications received; in 2017, the media outlet *Curbed NY* reported that eighty-seven thousand applications had been submitted for just 104 units.[77]

MOIA went on to list Housing Connect as an option for affordable housing for immigrants, right below a section notifying people of their rights concerning housing discrimination. The notification stated:

> Landlords cannot: Refuse to rent an apartment because of someone's immigration status, nationality, or religious beliefs; Post advertisements stating that certain types of tenants, such as immigrants or people from certain countries, are unwelcome; Fail to make adequate repairs or provide equal services to tenants because of their immigration status, nationality, or religious beliefs.[78]

While there are no official restrictions on noncitizens, the income ceilings—and assumptions of household size built into the criteria—have been severely prohibitive of applicants from immigrant backgrounds. The previous requirement of positive credit history along with social security or tax ID number added further barriers to immigrant populations. Although this requirement has since been revised to a twelve-month rental history instead, newcomers often rely on informal housing options to get by in their initial period in the city, for which they may not be able to provide legitimate proof of rent.

In my conversation with them, HPD officials maintained that these circumstances are assessed on a case-by-case basis and that they welcome alternative forms of proof that meet the "spirit" of what is necessary.[79] My informant emphasized the department's outreach efforts, like using housing ambassadors and hosting "Housing Lotteries 101" sessions, to increase accessibility for immigrants. But my conversations with the HPD came to a halt following a long quest for access to data showing the income brackets of those who had been successfully housed through Housing Connect. After a series of back-and-forth emails asking for proof on how the program benefited poor and/or immigrant communities alike, communication ceased altogether. It became apparent to me that despite its expressed willingness, the HPD did not want this information published in any way, which I found suspicious. While the Housing Connect portal has generated greater access to the application system, its distributive logics around housing remain intact. And so it remains entirely unclear if the process has had any positive impact on immigrant communities, who continue to struggle with it. Yunus tells me, "It's hard. There's no refugee coming to New York who's going to survive in housing. You're gonna work hard, but two-thirds of your salary goes in housing and you will not make it. It's hard."[80] Housing Connect demonstrates the fallacy of the insistence on "access" among technology advocates. Without an intentional structural and material reconfiguration of the dynamics that underpin the HPD's housing lottery, housing, and rent-seeking practices in New York in general, tech interventions like Housing Connect simply reproduce existing dynamics.

David had warned me that MOIA doesn't have the cultural competency to understand why clients don't want to cooperate or use its services. Officials will pride themselves on their progressive policies and the city's sanctuary status yet enable deportations based on benign welfare cases; deliver undocumented immigrants to ICE through an ill-designed ID program; unleash a data-harvesting Wi-Fi infrastructure with little to no safeguards; and reproduce inaccessible housing through digital interventions. "And then they wonder why immigrants don't cooperate."[81]

NAVIGATING REFUGE IN THE DIGITAL ANTISANCTUARY

New York's antianctuary plays out through informational precarity that disincentivizes affected populations from engaging with city services, and through the instrumentalization of the smart urban milieu for immigration enforcement. Consequently, the already limited ability of people to move freely in the city is under significant threat. In this way, the so-called border wall is survived by a far more expansive urban containment infrastructure that performs violence across racialized and gendered lines while soliciting self-disclosure, self-censorship, and self-discipline.

The unspoken consensus between the Department of State, the Mayor's Office, and Silicon Valley giants has cultivated the conditions under which digital forms of disaggregated urban migration control are possible. MOIA's digital pathways to sanctuary have been weaponized by law enforcement agencies, including ICE, and taken advantage of by other adversarial actors. However, these pathways have also given impetus to both a logic of externally imposed enclosure and self-imposed urban concealment. This chapter has referred to information control architecture and technologies that categorize precarious migrant populations as needy, criminal, and elusive, and contain them by limiting mobility and access through information panics. The enclosure of the digital periphery, in other words, continues to work across techno-developmental, -spatial, and -governmental lines.

But it is also simultaneously resisted. Marginalized migrant communities, activists, and allies are organizing apace with new realities. One of my Reset informants who supported asylees in their career development had noticed that her clients were increasingly "lying low," digitally.[82] The care with which data trails were avoided appeared as a consistent theme throughout my engagements, not just at Reset. This, paired with a distrust of mainstream sources of information (including large media outlets and the city directly), had led individuals with precarious immigration statuses to become "avid fact-checker[s]."[83] Farukh and many others at the refugee and asylum-seeker support group have a skepticism "for things that are not hard to obtain, including information." For Farukh, the information landscape had become too "political—there is a lot of anxiety and pressure among seekers of asylum, who worry about their residency status."[84]

At the same meeting, Abbas and Talal explained that there are two ways to securely gather and share useful information: through a diaspora-based WhatsApp group, referencing their membership in Guinean and Sudanese groups, where news about immigration in the United States, as well as politics in Guinea and Sudan, were shared; or via US-based diaspora radio hotlines, which broadcast news about Guinea and Sudan to the United States and operate a phone line connecting listeners to corresponding radio channels in Guinea or Sudan to relay important news.[85] Both Abbas and Talal showed me their correspondence with the hotline, which was coded as a favorite contact on their phones. Precarious immigrant communities go to painstaking lengths to keep the information environment decentralized; this was as true for Farukh, Abbas, and Talal as it was for Grisha and his clients at RUSA LGBTQ+ who engaged in "kitchen talk" as an antisurveillance tactic:

> So, you get together with friends in your kitchen, you're smoking, drinking or whatever. And you're discussing everything in the kitchen. It's like . . . from Soviet [times], because everybody [knew] big brother watched over you . . . that's why you will go through the kitchen. Because the kitchen didn't have surveillance. And you can discard the politics and criticize the government.

Grisha underscores how fear cuts through all aspects of socioeconomic survival in the city. Whether you're looking for information about or access to immigration services, housing, or work, there's a widespread abstention from engaging with the city's digital offerings. Instead, everything—from immigration information, access to services, and employment—is mostly "word of mouth because we cannot go and like [*scream*] 'Hire us!'" Grisha refers to kitchen talk as a contemporary version of an informational underground railroad: "But word of mouth, because it's face-to-face, it's more important in our culture. It's more secure."[86] Through such practices, newcomers learn how to navigate the digital antisanctuary.

At Reset, the decentralized approach was woven into the social fabric of the workplace. In my early engagement with Reset, Anita had insisted we needed a better understanding of where we might find housing for asylees and refugees on special visas, as she was convinced no one at Reset knew. Yet, my interactions with caseworkers—who themselves had been refugees once—demonstrated that this knowledge was very much shared between caseworkers and clients from their respective diasporas. For Albanians, it was the old Vatra Albanian American society and Alb TV, while the youth engaged mostly on the "Albanian Roots" Facebook group. For the Russian and formerly Soviet diaspora, it was the *Ruske Reklama* newspaper and kitchen talk; for Sudanese people, it was WhatsApp and, in some neighborhoods, an internal Wi-Fi mesh network that extended several blocks, providing free connectivity and an intranet for information relevant to the community. While this made it harder for Anita to centralize knowledge around viable avenues for housing, she admitted that "the division is both healthy and inefficient." Healthy inefficiency stands in stark contrast with the efficiency-maximizing principles that digitalization, surveillance, and automation often posit. It also denotes a subtle but important resistance to the centralization of intelligence on vulnerable diaspora networks, breaking with implicit surveillance norms and centering existing community practices.

Anneliese and Erin from Human Rights First insisted on the importance of real face and body networks. This is why they started running the refugee and asylum-seeker support groups, creating space for sharing best practices around how to work the system: "They know this in a way [our] organization just doesn't."[87] These are spaces where best practices, solidarity and emotional venting occurs. These "underground world[s] of different ethnic communities" were organized to foster solidarity and a sense of belonging not otherwise provided by the city. Erin reflects on how she had been raised to know about these informational hubs in the context of the West African diaspora, given her Malian heritage; Annaliese nods, pointing to the Cuban networks she remains a part of.

Several months after returning from New York, I caught up with Anneliese online. This was amid a further intensification of ICE raids and the acceleration of Mijente's #NoTechForICE campaign, just as public awareness around ICE's entanglements with Silicon Valley giants had started to grow:

> It has really made people on guard in a way I haven't seen before. I think that's the intention, mission accomplished. It has made people less eager to apply for legal asylum. . . . Of course, people are still doing it, but people are worried. . . . It feels like a lot of the work we're doing is such a waste of time. We're having to put out so much fire! We should be out there working on cases, working on trauma. But it's by design; to distract us from our ability to do our work. Our clients are always traumatized, always adapting to a new culture, [always] rebuild[ing] their life; but now in addition to all those things, people also have to be afraid of dealing with ICE. . . . It comes up at every meeting. . . . What's gonna happen to the people around me? What's gonna happen to my family? . . . People are worried that having a pending asylum application won't necessarily keep them safe.[88]

The moment Annaliese describes is sensitive and complex, to say the least: a technologically sophisticated border enforcement regime acts as both a distraction *from* important immigration work and an active gaze *on* immigration work and life in the city. This duality further constrains and disciplines life in the digital antisanctuary. To challenge this would mean to dismantle the digital infrastructure of the city altogether. In the words of Amelia, "At the end of the day, infrastructure is infrastructure. Who owns the networks?"

In stark contrast with most city initiatives, Good Call NYC stood out to me. It is a simple call-based hotline that connects individuals to their nearest available public defender in less than forty seconds. Set up through a social incubation program at Blue Ridge Labs at the Robin Hood Foundation, Good Call was set up with the understanding that arrested individuals who did not connect with a legal professional within their first thirty minutes of being detained were more likely to be incarcerated on false grounds. The concept is simple. Individuals frequently don't know who to call upon being transferred to a precinct. Contacting family or friends is often not a source of remedy. Good Call's intervention asks that affected communities memorize the number 1-833-3-GOODCALL. Backed by Twilio (a cloud communications platform that allows for complex routing of calls and text messages), Good Call can put New Yorkers in touch with their nearest participating public defender. Think of it as an inverted 911 service for racially targeted populations. Through partnerships with the Bronx Defenders, the Legal Aid Society, Brooklyn Defender Services, Neighborhood Defender Service of Harlem, New York County Defender Services, and Queens Law Associates, the hotline covers all five boroughs of the city.

Rather than insisting on a smart, aesthetically sophisticated, high-tech product, Good Call appeared to focus its work on the augmentation of existing channels of resistance. I learned that the organization was founded and housed within one of the premier US-based philanthro-capitalist foundations and incubators. However, it struggled to receive funding beyond the initial incubation period due to the organization's refusal to act as a marketing platform, charge for, or benefit off of users in other ways.[89]

In many ways, Good Call's approach veers on the side of neo-Luddism (as far as dismantling the inequitable gains and losses of the rapid development and deployment of new technologies is concerned). Mark explains at our second meeting that Good Call has made a point of neither accessing nor facilitating access to data from clients using the services. Much like David from BDS, Good Call advises partnering lawyers to inform clients not to tell them more than they have to, as adversarial surveilling parties may be a concern. Unlike its city-run competitors, "we [Good Call] don't market for this [toward undocumented immigrant communities] as it can look suspicious and create unnecessary visibility [of them]," says Mark.[90] While Good Call was not built explicitly for undocumented immigrants, it is not a coincidence that the service has sought to rapidly scale to cater especially to communities affected by deportation raids.

New York, which has a long history of being an aggressive site of immigration enforcement, assessment, and control, became a digital anti-sanctuary long before Trump came to power, and this continues today, as the Biden administration's "smart border" policies start to take shape.[91] The border wall is but a symbolic front, a distraction from the internal process of bordering enabled through institutional decay and the digitally weaponized city.

The unspoken consensus between the DOS, MOIA, and Silicon Valley giants has led to the development of sophisticated digital urban infrastructures, priming the city for disaggregated migration control. These technology deployments, in turn, insert technology actors in the business of governance. While initiatives such as LinkNYC, IDNYC, and Housing Connect purportedly give the city's refugees, asylees, and undocumented immigrants access to crucial digital services intended to improve life in urban refuge, they also led to information panics among those communities. This happens when the cost associated with the usage or exposure to these services is perceived to imply volunteering data to aid in the marginalization of migrants.

In hostile immigration environments, the availability of a vast number of data points can be instrumentalized by federal immigration enforcement officers for location tracking, detention, and eventual deportation. The fear of deportation against the backdrop of an advancing digital urban infrastructure has exacerbated the information panic in New York, leading communities to engage in practices of refusal such as urban concealment. Despite this, cities such as New York continue posturing as sanctuaries, while facilitating the rapid and lucrative entrenchment of Silicon Valley in the fabric of urban governance. The city is, in other words, a legitimizing ground for the tech companies, which instrumentalize it for the extraction of techno-racial capital. Technology giants superimpose neocolonial relations of power in the urban context, while marginalized populations play the role of test population, conscripted into the digital periphery. Under the auspices of servicing these communities, technology corporations can win major contracts with cities that not only finance their interventions, but legitimize their access to population data, which is in turn commodified and used for the iteration of further products.

5

DIGITAL REFUGEENESS IN BERLIN

IN THIS CHAPTER, I investigate refugee tech in Berlin. Contrary to New York City, these interventions do not necessarily sustain surveillance structures related to immigration enforcement in the city; nevertheless, they commodify and datafy subjectivities. The increased availability and usage of apps to access information, services, work, and housing potentially transforms the ways refugees access life in the city. With little to no oversight and accountability, this neoliberal approach to urban refuge in Berlin risks perpetuating the deep-seated myth that refugees and vulnerable migrant populations are made of fundamentally different matter—that their needs, literacy, and even desires around flexibility and stability are distinct and by nature more precarious. This, in turn, gives rise to *digital refugeeness*. Digital refugeeness exists for the enjoyment of experimentalism and solicitation of technical and financial capital between urban entrepreneurs, larger tech companies, and governmental as well as nongovernmental institutions, rather than demonstrating relevance or benefit for the communities it encompasses. While in New York, the digital antisanctuary was fundamentally about disciplining physical movement, digital refugeeness is about exploiting the subjectivities of moving bodies. These encounters continue to shed light on the workings of the digital periphery, and how seemingly disparate and decentralized forms of digital socioeconomic interventions convert refugees and their conditions into a laboratory for the extraction of capital.

SITUATING THE REFUGEE IN REFUGEENESS

NADIM: Like, I certainly lost a couple of friends or contacts to that behavior. Because, yeah, of course, people don't expect a refugee to come and say, "You're full of shit" or "Your idea doesn't work."

MATT: Why not? Why shouldn't a refugee?

NADIM: Because you just get help. That's it. Like, that's . . . that's the thing. And well, if we ask for your help or to volunteer, [you're going to do it] probably because we're gonna frame [your project] for the media.[1]

Nadim is referring to the boom in digital refugee initiatives—especially apps—that Berlin experienced in the two years prior to the 2016 EU–Turkey deal (intended to restrict the number of asylum-seekers coming into Europe).[2] In my first few weeks in Berlin in late February 2019, I had conducted a mapping exercise in collaboration with the Betterplace Lab, which found at least seventy such initiatives, with a large proportion of them having been established in 2015–16. With self-identified remits to address issues as diverse as orientation information dissemination for new arrivals to job-matching, language learning, and housing, these technologies seemingly offered to extend access to integration for newcomers.

The initiatives were driven in large part by a combination of volunteers, technologists, and civil society organizations. Between the second half of 2015 and the second half of 2016, projects were emerging weekly, at times almost daily, before tapering off in the latter half of 2016.[3] The sudden decrease in initiatives is attributed to "a dimming of public and media attention to the issue and a decrease in the number of arrivals"—again, due in part to the EU–Turkey deal of March 2016.[4] As the narrative of *Willkommenskultur* ebbed, the field of refugee tech consolidated; the tools that remain have tended to be propped up either by municipal governance (e.g., Integreat and Handbook Germany) or by big tech corporate social responsibility strategies (e.g., Salesforce's investment in Jobs4Refugees, Google's investment in Singa and Techfugees, and Meta/Facebook's investment in ReDi School).

By the end of my fieldwork in the summer of 2019, around forty refugee-tech initiatives remained. Nadim, whom I met during a refugee-led research workshop organized by a group known as G100 Berlin, had been contracted to work on an earlier mapping project with Betterplace before my arrival. He's acutely familiar with most of the initiatives, not least as they appear to reproduce a similar tradition of techno-humanitarian interventions that he encountered while living in Greece. He indulged me in a series of frank conversations surrounding the politics of Berlin's refugee-tech industry. Nadim is one among a dozen newcomers and former refugees I came to know during my time in Berlin, most of whom largely hailed from advocacy, resource, and information-sharing organizations such as the G100, Sharehouse Berlin, and Syrische Frauen in Deutschland. While few were tech entrepreneurs themselves, I spent a large proportion of my time engaging with refugee-tech initiatives led by predominantly German and European individuals and the organizations that gave rise to them, including Techfugees, Integreat, and Jobs4Refugees. Through these encounters, I came to understand Berlin's digital refugee response as a form of digital ghettoization, in which the abstraction of digital refugeeness is constructed as an avatar to attract funding and claim political solidarity. While chapter 4 explored how digital initiatives transformed the urban environment for people with transient immigration status in New York, this chapter is a foray into how refugee subjectivity in Berlin is transformed for capital by tech.

Critical techno-cultural inquiry into how refugees and newcomers use (or don't use) these technologies, and how they are experienced,[5] provides invaluable insight into the political workings of purportedly apolitical and benevolent tech interventions. Contrary to New York, refugee-tech in Berlin does not necessarily sustain surveillance structures related to immigration enforcement in the city; nevertheless, the increased availability and usage of apps to access information, services, work, and housing transforms the ways refugees access socioeconomic life. Refuge in Berlin today is, unavoidably, digitally mediated, constrained and enabled by digital maps and apps. In the name of techno-urban entrepreneurship, spheres under the purview of local authorities and the state have increasingly been encased in privatized digital layers specifically tailored toward refugees.

Digital refugeeness allows for the development of refugee tech without the input or involvement of displaced individuals at the iteration phase. Despite the broad use of existing channels of both digital and nondigital communication among refugees and newcomers, digital refugeeness mutes the subjects of concern by substituting refugee leadership in assuming their backwardness based on tired tropes of technological inferiority. Mainstream channels are ignored and sidelined in favor of the development of new platforms altogether. Digital refugeeness allows for the movement of capital between urban entrepreneurs, tech companies, and governments, by utilizing the racially marginalized and "needy" imagery of refugees.

The digital periphery disciplines refugees with the language of gratitude and presumed backwardness, creating a regulatory vacuum in which those affected by governance are unable to change it; they are flattened, instead, into digital refugeeness.

PACKAGING THE PROXIMATE OTHER: THERE'S AN APP FOR THAT

At the time of commencing my research in Berlin, four years had passed since the death of Alan Kurdi had sparked significant outrage across Europe's liberal artists, policymakers, and civil society alike.[6] Germany's "refugees welcome" sentiment had grown in tandem with the rise of Europe's "new radical right,"[7] as the continent sought to reckon with its image as, at best, passive and, at worst, permissive concerning the continued violence and death experienced by displaced populations attempting to reach its shores. Reflected in these supposedly affirmative approaches to refugees, however, is a continual process of reproducing the same logics underpinning anti-immigrant sentiment. The migration scholar Ida Danewid explains how the much-celebrated *Willkommenskultur* (culture of welcome)

> reproduce[s] the underlying assumptions of the far right: Namely, that migrants are "strangers," "charitable subjects," and "uninvited guests." By focusing on abstract—as opposed to historical—humanity, they contribute to an ideological formation that erases history and undoes the "umbilical cord" that links Europe and the migrants that are trying to enter the continent.[8]

Europe's renegotiation of itself as a liberal progressive project is not a new phenomenon. However, it has been more of a project of historical obfuscation. This approach curates an artificial European civic nationalism supposedly grounded in a collective vision of peace and prosperity. It forecloses the centrality of the continent to the production of racial capitalism, as explored in chapter 1—and in so doing, the genealogy of contemporary institutions and structures, including that of the European Border and Coast Guard Agency (Frontex) and the multiplicity of strict and violent post-9/11 national border-control regimes. These cannot be disentangled from the continent's historical production and reproduction of racial order.[9]

Refugee tech—apps in this case—serve a similar function to *Willkommenskultur.* In so far as they mirror the German gaze, they fuse a plethora of diverse displaced individuals to the designation of the dehistoricized, decontextualized refugee. Nadim, a twenty-eight-year-old Syrian researcher I met in Berlin, deconstructs the underlying premise of refugee tech and the people behind it, unveiling a fundamentally essentializing process:

> The thing that bothered me the most is people who think they're helping, but they're not. And they're getting all the credit for it. . . . So like, okay, let's say somebody came up with . . . an AR school for refugees. Like augmented reality, and everybody would be wearing the 3D thing and like, learn languages from [it]. . . . Yeah, it sounds cool, right? Sounds like futuristic and all. I mean, you might get funds, it might be framed and covered on a couple of TV channels or newspapers. But yeah, that's all good. But this is *for you,* so does it really work [for refugees]? That's the question. And it seemed like many people were not really asking themselves this question. . . . That didn't really go well with me. Because [in tech development], most probably they are actually treating refugees as an object—as a very fixed, well-known object, which is not the case.[10]

The branding of digital initiatives in the name of refugees (e.g., Integreat, Jobs4Refugees, Refugees Welcome, Refugee Text, Techfugees, etc.) is too specific and lacks nuance, context awareness, and recognition of the diversity of people, experiences, and backgrounds that make up refugees. This allows the already racialized figure of the refugee to take on a technologically mediated managerial disposition of digital refugeeness.

Nadim and I sit in a café in Friedrichshain in East Berlin. Since the fall of the Wall, this formerly working-class area has become established as an up-and-coming, young, and at times radical stronghold of artists and creatives. With its artisan market, famous nightclubs, anarchist squatters, and Australian cafés, Friedrichshain stands out from its shared borough, especially the nearby "Kreutzkölln" district. Kreutzkölln and the surrounding area was home to many newcomers, with Arab, Turkish, Syrian, Lebanese, and Persian restaurants, grocery stores and residents. Friedrichshain, on the other hand, was the liberal center of Berlin's well-meaning white-saviorism. From this safe distance, NGO workers and technologists with messiah complexes could dream up their sociotechnical visions for saving refugees through their various integration projects. Nadim and I had met a few days earlier, when we attended a workshop organized by a newcomer research initiative known as the G100, in response to the seeming lack of consultation and inclusion of refugees and migrants' perspectives in policymaking and integration interventions at large. Nadim had appeared ambivalent about the workshop, cautioning that even with its valuable insights nobody would listen.

Alongside my conversations with newcomers, I had long-term conversations with technologists engaged in the refugee-tech space, like those at Techfugees. In a conversation some weeks before the workshop, a Berlin chapter lead of Techfugees had gone to painstaking lengths to pitch me his vision of a solution to the refugee problem: "Germany is not doing enough at a policy level. Refugee integration should be offered as a service."[11] Matteo's vision is one in which municipalities and venture capitalists fund third-party private organizations to provide tools for "refugee integration, [where there is already] a war for funding." Matteo explained that his approach was "apolitical, and therefore better for the state. It would [help the state avoid] having to wake the AfD beast."[12]

Matteo's perspective is not just a vision, but an astute reflection of how epistemologies of survival are sidelined in favor of capital. It's a reminder that well-connected European urban entrepreneurs like Matteo are more likely to receive funding, support, and an audience for their interventions in refugee governance, while individuals like Nadim, with lived experience and membership in the communities in question, are asked to simply be grateful for the ideas of the Matteos of the world. Matteo laid out his four-point plan: "First, we set up a syndicate of VCs and funders across Europe. Step two, we hook up tech entrepreneurs around Europe with funders—also because refugees don't stay in one place. I want to create a sharing house of ideas for refugees. Finally, we would sandbox different configurations of tools." Throughout my many engagements with Matteo and Techfugees Berlin, he remains resolute in his primary objective to woo venture capitalists and entrepreneurs building tools for refugees. On one or two occasions, he did suggest and contemplate the possibility of funding a tech incubation program for refugee women. But he remained predominantly concerned with (1) establishing Techfugees as a clearinghouse for VCs interested in putting money into refugee tech and (2) continuing running hackathons that would help solve the refugee crisis.

One sympathetic Techfugees Berlin member and collaborator of mine, who had positioned themselves as a critical check and balance on the organization's techno-solutionist orientation, describes the group's approach as voyeuristic at best: "There's some real camp voyeurism on that WhatsApp chat. People just want to get to the [urban] camps to find needs for their solutions."[13] Recasting refugees into an essentialized other, the refugee tech enterprise transmutes human subjectivity into financially decipherable objects, in the process attracting capital from tech giants, venture capitalists, and local authorities. Whiteness is more receptive to one-dimensional, decontextualized "refugees from refugee land," than to equally complex human beings. Nadim shrugs and says, "And now [they're treating them like] they're all in one camp, and they all are gonna be the same."[14]

THE REFUGEE-TECH BUBBLE

The president of the Federal Office for Migration and Refugees expressed concern in November 2019 that "migrants and refugees who work in very low paid jobs are stuck in very precarious situations and could suffer from old-age poverty in the future."[15] The president's concerns came out against the backdrop of one of the highest influxes of newcomers since the postwar era, all of whom needed to be processed into German society. In response to this, the common justification became somewhat focused on refugees as economic stimuli, in a moment of an aging and depreciating labor force. It is unsurprising, then, that many flexibilized work initiatives emerged, purporting to tackle the issue of replenishing the German labor supply, with potentially precarious results for refugees. When held up against the consistent complaint from newcomer communities about the barriers to having their qualifications accreditations recognized, it is easy to see how highly educated individuals fall into a precarious work trap. What is more, the German refugee-tech scene tends to discursively describe two forms of refugees: the seeker of *any* job (usually menial/flexibilized) and the tech entrepreneur (looking for coding lessons and funding for their tech project). Both scenarios are unstable and essentialize refugees into a form of neoliberal conformist urban-dweller.

"A business that's getting value because it's refugee business. . . . Of course it's good, but a failure of the system," Sarina tells me.[16] As one of the G100 organizers, she had initially invited me to join as a notetaker and support one of their workshop facilitators. The workshop itself was geared around engaging with newcomers of varying immigration statuses, to learn about challenges that were not currently addressed by interventions made available by the state (these included civic engagement, access to the job market, access to education, and personal development).[17] Sarina explained that the G100 had become an important space for newcomers because Germany had fallen short of recognizing the individual needs of refugees and had instead created a generic and homogenizing integration pipeline. While Sarina had been completing her degree in medicine in Syria, she realized she would have no say in what job she would end up with upon arrival in Germany: "They decided it for me and transferred me to the job [at the job center]." As a result, she became a translator, recounting that it was considered one of the positions with greater mobility, but that ultimately this outcome was arbitrary as the job center only considered language among her potential employable assets.

Despite these low expectations, Sarina went to great lengths to ensure she would be able to requalify as a doctor, even if it involved restarting her degree. Sarina described how many others in her circumstances had been offered to start a "refugee business," be it a catering house or a café, but that ultimately the substance didn't matter as long as the refugee narrative was being sold. "There's a difference between selling tools made by 'victimized people' versus selling products made by professionals whose craft it is. If you take away the victimized narrative, it's not special." In her experience, Syrians in Berlin are converted from agricultural engineers to Uber drivers, from lawyers to SUPERMARKT workers. Constructing and selling the proximate other strips individuals of their discernible characteristics and flattens them.[18] The few digital initiatives Sarina had come across added marginal value to newcomer life in Berlin, as "mostly they repost existing job ads on their own platforms. The only value add is that they might call the place you applied to, to see whether they considered you."[19]

Online representations of race are culminations of programmed whiteness and offline perceptions of race.[20] In this case, online and offline iterations of refugeeness come together to solidify the purportedly known nature of the proximate other. The following examples are particularly illustrative of this paradigm.

Integreat

Integreat was born out of efforts by volunteers who were helping newcomers in the wake of the 2015 arrivals. The team was situated in different volunteering organizations that were all facing the same problems: handing out printed flyers with information to help navigate the immigration bureaucracy did not make sense when laws and regulations were changing every week. Camps were emerging and then closing in many parts of the country. A lot of information quickly became obsolete or was simply not communicated. The Integreat team got together around the idea of creating an electronic platform of sorts—in this case, an app and website—where up-to-date information could be provided digitally for refugees and migrants. Integreat, available on Android and iOS and as a web app, works through contracting with different municipalities across Germany that wish to provide digital information about their bureaucracy to refugees and migrants. The Integreat team informed me that the municipalities started using the tool—which has a user-friendly municipality-facing dashboard that government workers and the project team can populate with information—in place of a municipal website, with a particular focus on EU migrants.

As municipalities are responsible for updating the information on the platform and do not necessarily have refugees in mind as their first service population, information quality varies widely. Similarly, while the project received its initial round of funding largely in response to meeting the challenge of information provision specifically for refugees, Integreat employees currently do not have a definitive method by which they understand who uses the service. As it stands, they rely on word of mouth from city authorities, with whom they also conduct their usability testing. However, they have no awareness of how refugees or newcomers use the tool. The Integreat app, in other words, runs the risk of becoming a symbolic front for action on the part of municipalities, without any real obligation to ensure that crucial information is delivered or understood.

Jobs4Refugees

Among a host of other digital job-matching services, Jobs4Refugees generated particular elation in Berlin. It not only provides a platform through which job-matching could occur but also, to varying degrees, facilitates the negotiation process between employer and employee, typically in an attempt to "warm them up" to the idea of hiring a refugee. While the initiative is novel, conversations with the team revealed that the longest recorded time of someone staying in a Jobs4Refugees mediated position has been around six months, and that this tended to vary greatly.

Jobs4Refugees also partners with corporate giants such as Facebook, Salesforce, and Accenture. It is worth noting that these big companies involved with developing refugee-oriented technologies have been documented to have bad data-protection practices. In some cases, they have contributed directly to the surveillance of migrants by actors in the deportation machine. For instance, Salesforce has contracted with ICE, and Accenture spearheaded the ID2020 Digital Impact Alliance's blockchain-based ID system. Furthermore, it is unclear what client data Salesforce—which provides the dashboard for communication between Jobs4Refugees and its clients, free of charge—has access to. There are many initiatives like Jobs4Refugees (see for instance Workeer), some of which more openly offer flexible/remote or microtask-based digital work. These tools, with large technology companies as their backers and precarious labor practices, risk encoding urban precarity among newcomer populations, while keeping capital in circulation and generating potential data dividends for backers such as Salesforce.

Handbook Germany

Funded directly by Bundesamt für Migration und Flüchtlinge (BAMF), and in partnership with T-Mobile and Adobe, Handbook Germany has in recent years attempted to position itself as the go-to source of information for newcomers. As per their description:

> Every country has its specific characteristics. We know what it feels like to be new in Germany because many of us have gone through the same experience. Here we offer essential tips on asylum, housing, health, kindergarten, university, work, and much much more in seven languages. The Handbook Germany team provides crucial information on an extensive range of topics—here you can find from A to Z of life in Germany on a single website![21]

Contrary to many information dissemination tools in Germany, Handbook Germany has dedicated editorial staff who coordinate with BAMF to deliver important information through text, short videos, and Facebook messages. Yet, there is still little clarity on how much refugees and newcomers use the tool.

"You download them, and then they're incompatible, or they stop being serviced after a while. Most of the time, if people do download a refugee-specific app, they end up deleting them—but technologists use the download count to justify continuing anyway," says Hakan, a researcher and Syrian refugee I met at Sharehouse Berlin.[22] Hakan, in his best diplomatic manner, confronts the uniquely European phenomenon of being surprised that refugees have smartphones, a sentiment emerging from the conflation of refugees from Global Majority countries with poverty, depravity, and the underclass.[23] He shrugs, recounting how numerous European researchers had approached the newcomer residents of Sharehouse Berlin to study their everyday smartphone usage. "We've done this research already," Hakan insists, while searching his laptop for a paper he coauthored with the Berlin-based academic Safa'a AbuJarour. AbuJarour and Hakan had analyzed the mobile phone home screens of 101 refugees in Berlin (subsequently compared to home screens from 107 Germans and 72 immigrants), which revealed what many in Europe might have considered a surprising result: more often than not, refugees rely on similar apps for obtaining information as anybody else. WhatsApp takes the lead, appearing on 69.64 percent of all phones, followed by Facebook Messenger at 37.86

percent, YouTube at 37.50 percent, Facebook at 31.79 percent, Instagram at 26.79 percent, Telekom.de at 20.71 percent, DB Navigator at 16.79 percent, Arabdict at 13.93 percent, Spotify at 13.21 percent, and Snapchat at 13.21 percent.[24] Mainstream tools with household names were far more frequent on home screens than anything designed specifically for refugees.

A close facilitator and collaborator during my time in Berlin, Edin (also a mutual connection of Hakan's) had aired his grievances on this subject with me just days before Hakan:

> It's like they think we are lab rats. They do a big survey, collect our information, and then they leave, and we never hear from them. At first, we were happy to help because we thought [these] Germans could help us, but after a while it [got] exhausting. I personally have been involved in twenty projects at least.[25]

A common thread across Nadim, Sarina, Hakan, and Edin's grievances is the extractive relations between refugee-tech initiatives and the refugeeness of newcomers. AbuJarour's work reveals that refugees and newcomers hardly use this refugee tech. Nevertheless, the salience of refugee tech in attracting funding—in addition to its discursive power—is not a trivial nor immaterial disruption to newcomer life in Berlin. Myria Georgiou incisively observes that "not everyone speaks and is heard in the same way; not everyone is equally represented, even if most are digitally present" in the digital age.[26] In other words, these initiatives rely on newcomer communities, superficially, to symbolically legitimize the existence of refugee tech. They rely, desperately, on the image of refugees, not their voice, which reinforces iconographies about the refugee that sustain a loop that keeps funding out of newcomer communities and in the hands of technologists. This is why Nadim calls refugee tech "a bubble":

> Helping people should not be a wave; it should not be a bubble. Now, helping refugees *through* tech is even worse, and should not be a bubble. Now

> entering the tech scene to help refugees and get plenty [of funding] . . . it should not be something "cool" to do. . . . There's a lot of the application where they're like, "Let's do another WhatsApp for refugees." . . . Like, well, why don't we just use WhatsApp for refugees? "Because refugees need to use another application, wow wow wow, you didn't know that?" So yeah, "Let's make a new social network for refugees." . . . I see a lot of this![27]

The generation of value for these initiatives, in other words, takes place outside the tech company. In her work on postindustrial value creation in the digital age, Terranova describes how social, cultural, and economic networks that surround and exceed the internet, what she calls the "outernet," "connects [the internet] to larger flows of labour, culture, and power."[28] Even though the value of these tools is essentially negligible for newcomer communities, the whiteness underpinning the internet has given way to a modus operandi of value generation that is chiefly concerned with reinforcing and selling German generosity through the proximate other, rather than servicing diverse problems.

VALORIZING NUMBERS

To carry Nadim's bubble analogy to its conclusion, the politics of refugee welcome and integration is ontologically intertwined with not only racial logics in general but the racial logics underpinning Silicon Valley in particular. Technologists often claim that refugee-tech apps provide "a free technology for social relations." In reality, the apps' value is only possible because technologists sit on the capital means to "enclose technology and social relations [previously] in the commons."[29]

While the tech industry's investment into refugee-tech apps should raise concern, the diversion of funds to refugee tech, from the likes of BAMF, German municipalities, and the UNHCR, is of even more alarming. In the same way that 40 percent of European start-ups that are classified as AI companies do not, in fact, use artificial intelligence in a way that is "material to their businesses," refugeeness has also become "catnip to investors."[30] In contexts of migration, technology often acts as a gateway for authorities and institutions to appeal to narratives of gratitude and victimhood in refugees for political expedience. It advances an image that positions them (aid recipients) as fundamentally different from us (aid providers).[31] This is particularly problematic against the backdrop of a hostile global immigration environment,[32] in which the politics of welfare state membership is inextricably tied up with xenophobic posturing by right-wing parties, as evidenced by Matteo's comments on "refugee integration . . . as a service." Hidden behind the outward-facing veil of welcome is a techno-assimilationist fantasy of control. Control that both circulates capital and designates the racial order of refugeeness.

After some months of chasing, I was able to secure a sit-down with a leader at Jobs4Refugees. We had been communicating via calls and chat for a while and had planned to meet for a more detailed conversation around the inner workings of the initiative. Tom tells me things have been hectic around the office, with the team having to prepare for the visit of Prince Charles and the Duchess of Cornwall.[33] He tells me the project came about when they learned they could register nearly two hundred refugees in two days just by hanging up signs in different languages near urban camps, stating "we're trying to help you find work":

> We quite simply start cold-calling employers asking them whether they're open to hiring refugees, and then facilitated between the two sides. And after one month, we thought, this is perfectly okay. This is actually leading somewhere that we feel okay about. If us two numbnuts can do that, then, then anyone else can do that. . . . We had quite some media coverage in the beginning, which led them to some people actually phoning us with more money.[34]

Tom explains that they used the money to set up Jobs4Refugees as an organization. In its Facebook page likes, Jobs4Refugees found an unexpected source of symbolic capital. Tom conflates what visitors to the page would see as the aggregate number of likes as the number of refugees, stating that the user journey "starts on our Facebook page, where we have twenty-two thousand refugees." Based on Tom's responses and the fact that Facebook pages cannot verify the immigration status of their followers, it seems that Jobs4Refugees is not actually interested in quantifying refugee engagement. I ask Tom what, in his experience, attracts refugees to Jobs4Refugees, as opposed to other initiatives or going through anchor communities, other job boards, or the government:

> To be honest, I don't know. We've never had a problem in admissions, we worked with a bunch of migrants. And we went through Facebook, we hardly ever had a problem with the acquisition of refugees or like getting a recognition for a job. So . . . it was never something that was kind of like *a highly effective cost to evaluate.*

One can speculate that for Tom, determining why someone would use Jobs4Refugees is not "a highly effective cost to evaluate," because it ultimately remains a secondary objective. At the time of our conversation, Tom and his team had recently courted Salesforce as a partner. With a prospect of significant funding from the Salesforce Foundation, Tom admits that there is "a tension between also aiming for funding" and continuing the work they had done during their pilot.[35]

At the time of writing, Jobs4Refugees has partnered with both Facebook and Accenture and won the German "integration prize." I asked Tom how long they were projecting to be in business, expecting that a refugee job-matching service would either dissipate as the number of newcomers tapered off or be absorbed by the state: "I guess there's no like . . . there's no expiry date in mind."[36] Few newcomers feel the material impact of the existence of German tech entrepreneurs who work from trendy offices at the Impact Hub (just a short walk west of the emergency refugee camps at Tempelhofer Feld), profiting on the back of overblown Facebook likes, royal visits, and big tech investments.

This was not the first time I had heard about a tech initiative selling refugee demographics to funders. Alongside my research in Berlin, I was in frequent conversation with informants from the humanitarian sector who had previously worked on the Signpost and Refugee.info initiatives (information-sharing projects that used to be jointly held by the International Rescue Committee and Mercy Corps). They similarly claimed that the "1.5 million unique users in 8 different languages" reported by a Mercy Corps employee as the number of refugees the initiative engaged with,[37] was actually the total number of followers across a handful of different Refugee.info Facebook page instances.[38] Initiatives such as Jobs4Refugees and Refugee.info demonstrate that value can be extracted (or made up) from refugeeness alone. As per Terranova, tech intervention in and of itself cannot mobilize capital without the cultural and political significance bestowed on by it by the images of the "vulnerable" refugee and "benevolent" funders.

VALORIZING FAILURE

Another modality through which refugee tech functions is failure. We would be remiss to brush off initiatives that fail as inconsequential to the materiality of refugeeness. The idea of refugee identification through blockchain has long captured the imagination of technologies, well before ID2020 and World ID. Taqanu is one such initiative that provides some insight into this. Yanos, a representative of Taqanu, tells me that the blockchain-based digital banking and identification system was conceived

> four years ago, when Hungary built a fence against refugees during the heat of the so-called refugee crisis, and I was living up in Norway. . . . And I felt like I could . . . should do something maybe and having a couple of

> conversations with refugees who I could meet in Norway, it kind of became fairly clear that like, their largest issue is accessing a bank card or bank accounts. So, then the idea came, like let's make a bank for refugees.[39]

He says that the idea was essentially to use "social credentials and user behavior" as an additional layer of authentication, in the absence of trust in printed paperwork. He uses the terms "untrusted network" to describe what he's trying to build, though it might have been more appropriate to call it "a network of the untrustworthy." However, Yanos had only engaged with refugees anecdotally. In an article I read before our interview, he expressed that access to refugees was one of the biggest obstacles to innovations like his.[40] Yanos tells me this was no longer an issue, as Taqanu had been brought in under several large partnerships, predominantly in Kenya, Uganda and Ethiopia, under the Smart Communities Coalition funded by Mastercard.[41]

Taqanu eventually ran out of funding and never fully got off the ground. In Yanos's view, however, it still played out well for Taqanu compared to other "identity sort of companies that like, created a lot of buzz, raised money, and then they are nowhere to be found." Taqanu has pivoted to exclusively providing consulting services to institutions, including the UNHCR, based on its "expertise." He explains, rather matter-of-factly, that "we are just doing consulting on the topic of identity and the theme is a bit [more softly] connected [to the original project], because money needs to be made."

The examples above demonstrate that refugee tech is a superficial way of engaging with refugees. It is also an intimate and consequential method of capturing and reproducing refugeeness. This reveals that refugee tech is a frontier for techno-capitalist extraction. Refugeeness is but an avatar, a categorization that, in addition to commanding capital, broadcasts a representational message about the raciality of refugeeness, which serves a disciplinary function[42] insofar as it conditions society to perceive anyone loosely affiliated to the avatar *as* the avatar, proper.

CAN THE DIGITAL REFUGEE SPEAK? INVISIBILIZED EPISTEMOLOGIES OF SURVIVAL

During a call with a lawyer from the Gesellschaft für Freiheitsrechte (GFF) or the Society for Civil Rights—an organization founded in 2015 that works "to promote democracy and civil society, limit surveillance and digital scrutiny, and enforce equal rights and social participation for all people"—I find out that German border officers had been caught running analytics software on the devices of new arrivals, analyzing the entirety of the phone's system for markers of the user's country of origin (using a broad set of data, including apps, languages, and historical location data).[43] In 2017, BAMF introduced a policy that allows the office to "extract and analyze data from data carriers such as phones in order to check their owner's stated origin and identity" in the absence of a presentable passport.[44] The system, which generates a report from every instance of extraction, is accessible only to lawyers, not applicants.

The GFF's findings reveal that the practice is not only an extraordinary breach of privacy, but also almost entirely useless. Its report summarizes that 64 percent of cases contain no usable results, 34 percent confirm the origin and identity claims of the individuals, while only 2 percent contradict the applicant's claims. This is unsurprising, given that—as established previously by AbuJarour—the configuration and selection of apps on the phones of newcomers are practically indistinguishable from their German counterparts (perhaps except for the Arabic dictionary, which could well encompass a multitude of geographic origins). The BAMF practice is yet another example of the persisting myth of digital refugeeness as a discoverable inherent characteristic.

As explored in previous sections, digital refugeeness supports the myth that displaced individuals are not familiar with the technical know-how and products of the West, and are in need of specific platforms. This myth is espoused by technologists and policymakers who believe that displaced individuals are unfamiliar with existing popular social networking tools. They believe that the needs, literacy, and preferences for stable versus flexible jobs of displaced individuals are distinct from their Western home communities.

In this section, I continue to reflect on the critical perspectives of community leaders and researchers on the tech interventions that continue to insist on digitally fusing them to their so-called refugeeness. I also draw on research findings collated by newcomer communities through the G100 workshop in Berlin, which go some way in communicating the gaps as they find them in Germany's refugee politics (which remain unaddressed by every refugee-tech initiative I came across). Groups like G100 Berlin aim to generate a more vocal discourse around the policy changes that are required to make lives for refugees better, steering the conversation away from its somewhat persistent techno-fetishism.

Emphasizing the cognitive dissonance apparent in how technologists in host communities reduce, simplify, and essentialize newcomers for technical expedience, Nadim tells me:

> You need to really understand this fragmented, diverse, complex community of refugees. Or else, you just . . . like you can't. Well, first of all, you can't help at all, right? And even if you take a small fraction of it, like, you might not really answer their needs.

Nadim recalls his fascination with the world of start-ups as he indulges me in his previous ventures. He reminds me that, of course, Syrians also set up digital initiatives to help newcomers. Crucially, however, these tools (such as Make It German and Bureaucrazy) do not make representational claims on refugee identity. Rather, they point out gaps in the everyday experiences of newcomers and provide shortcuts around them. Nadim says, "Technology is cool . . . it's nice, but it is not magical, and digital solutions . . . they're not going to work for this. And they're not enough. You need social interaction. Human to human interaction." In Nadim's experience, the tools that have been the most useful for displaced populations, even for "illiterate people [who] had nothing to do with technology," were Facebook and WhatsApp, "because they need that to actually survive and connect with the friends and family for strategic things."[45] Hakan had similarly noted in an earlier conversation that the most useful technologies were infrastructural, emphasizing tools such as Facebook groups over WhatsApp: "When I came in 2014, I didn't need info made especially for refugees—I had people around to ask, and preferred it this way."[46]

Edin and I are having coffee in Charlottenburg when he tells me about the first time he met the executives from Techfugees: "They announced 210 initiatives! Apps . . . where you could look up services in German or Arabic. And people were not using it."[47] Surely one of the tools would have had some salience with a newcomer, but Edin is steadfast in his reply when he compares the apps to the equivalent of a random person putting information out into the ether:

EDIN: Because we didn't know about it, we don't trust information from the internet. Why should I trust [it]? And I mean, you have no idea [who has shared it]. Okay, I'll tell you something. If you open this [opens his phone's browser] and you find whatever information you would like, would you believe it?

I shake my head.

EDIN: Okay, so this is how it is for us. It didn't happen in our life that we trust, that I open [this device] and then I have [trustworthy] information.

MATT: Okay. But if you knew that, for example, a guy called Mohsen had put the information up . . .

EDIN: Yes. Yes. If Mohsen, the guy that I know, yes. Then that's another thing—random people putting information [out]—that's not social media. So, it's like googling and just seeing an article, that we didn't [even] read.

Edin takes out his phone and shows me a plethora of Syrian knowledge-exchange Facebook groups based in Germany. He also shows me his Facebook Messenger inbox, where friends are asking him about refugee and newcomer policies. Edin points out one friend, Mansour, who wanted to attend the G100 but was worried about his safety given his disability: "Someone like Mansour might reach out to me or others to ask that we [accompany] him as he might need protection."

Upon Edin's recommendation and an invitation from G100, I took part in the organization's inaugural Berlin workshop. Throughout the day, rotations were made by some forty people attending the workshop across four different tables, with each table tackling a different topic. I was stationed at the "civic engagement" table, where the elephant in the room was quickly revealed to be integration:

> I hate this word *integration*, I hear it in the news, in class, in the street. (*Workshop participant*)

> At this point, I want to say that the word *integration*, it doesn't work. Because it means to remove and to delete everything I know in my life, just so I can live here. So, if we can change this word *integration* to *inclusion*. (*Workshop participant*)

> The word *integration* is shit because it assumes that we are a homogenous whole who can be integrated through one-size-fits-all solution. So, a solution would be to challenge the way the whole society thinks and talks about refugees. This is a process, a discussion, not just a policy to be carried out. (*Workshop participant*)

> We fall into the trap of perceiving the refugee population as one fixed entity. It is not, it's very dynamic, it's very unique. Integration assumes that this is something fixed. (*Workshop participant*)

The workshop took a life of its own as the exchange and critical insights on refugee policies and interventions became an avenue for airing and addressing grievances. Chief among them was the implication of integration insofar as it constructs refugees as a homogenous whole (what I refer to as refugeeness) and, remedially, the need for political organization through coalition-building across immigrant communities of various statuses. Notably, these grievances illustrate the dearth of understanding, whether deliberate or inadvertent, among refugee-tech initiatives of the newcomer communities they purport to serve.

For Nadim, this was the primary purpose of G100. He described his deep wish that people would realize how much power they hold:

> In a few years, one million refugees will have voting power. This is very serious. And if the young G100 community organizers are too inexperienced, they might not be able to see past the smiles of politicians. We need to realize the power we hold and the power that we [will be able to use] in places like Berlin and Hamburg, where power is concentrated in parallel with how othered communities are concentrated.[48]

REFUSING REFUGEENESS: BREAKING THE MOLD VIA FEMINIST KNOWLEDGE PRODUCTION

Sarina, the organizer I met earlier, has seen a form of organizing that can cut through the insistence on refugeeness. "Oh, it's still a Facebook group," she says, "[but] think of the knowledge produced in this group and how many lives changed because of it and the good advice is given and so on." Sarina is speaking about Syrische Frauen in Deutschland, a Facebook group set up with the intention of sharing knowledge about navigating life in Germany among newcomer women. Sarina is sure that no German would take Jobs4Refugees seriously, despite it being nineteen thousand members strong (just three thousand short of the number of likes on its Facebook page), because it's on Facebook.[49] But

Syrische Frauen was a far cry from interventions I had come across before. Sarina tells me that women "archive, [write] topics, tag [posts], and so on. And they have house rules for communication. And in my opinion, the way the group started, and the way it's working now, they have worked really hard to create this communication culture, and it has become a very supportive group."

Sarina introduced me to the founder of the group. Herself a Syrian newcomer, Yara had conceived of the group when confronted with the need for a centralized but human knowledge repository as she was preparing her move. The group, which began in 2016, was exclusively run by and for women:

> My friends, who are men, they were saying—even Syrian men who have their own groups—they were coming to me [saying] "Yara, please ask this question for me, you women know how to talk about things better than us," so I use it [on their behalf] for my close ones.[50]

Sarina and Yara had both told me about countless other Facebook groups that ended up being co-opted largely by men, who tended to turn these spaces into toxic political battlegrounds. Yara told me:

> All the information is only from the members, from the women who are in the group, and they're really interacting in amazing ways, no one posts a question unanswered; this is impossible. All the questions have people to answer them. And without asking or encouraging them. Just, people got used to sharing. I want to share my experience; I want people to know about what I know about Germany, we just made it a good experience for all women in the group.[51]

She walks me through the group. The landing page is a feed containing a pinned post explaining the available tags (these include *education*, *housing*, *childcare*, *immigration*, and *job center*, among others). Yara and her handful of content moderators had gone through every question posted and added a unifying tag so that visitors could jump straight to posts relevant to their query. Yara also underscores that many decisions are mediated through the consultation of fellow women in the group, emphasizing that this was an act of empowerment, as opposed to information-seeking, per se:

> For example, there was a post, a young woman was asking, "Hey girls, do you think that if I graduated from the university at the age of thirty-five, it's still possible to get a job, or do you think it's better that I don't get a master's now and just look for a job?" She was talking about the age [as a barrier], and that was interesting because 90 percent of the women were like, "No, go for the studies, you will have a chance," and that was really nice.

As per Urszula Pruchniewska, without explicitly being feminist, spaces like Syrische Frauen create a "bordered" safe space, which keeps at bay not only toxicity from men within the diaspora but also the German gaze and its insistence on a linear, top-down integration policy.[52] The group has flown under the radar—not because of a larger conspiracy to keep it secret, but because information about it was disseminated by members of the diaspora, *for* the diaspora. Not unlike André Brock's observations on Black Twitter, as technology augmented spaces for displaced populations are less inundated by the sensationalism afforded to the notion of refugee-tech during its early days, so their relevance—albeit in the mundane form of the standard-issue Facebook group—become more salient.[53] The existence of this particular iteration of a women's only "e-diaspora" in relative anonymity repels the spectacle that would otherwise alert technologists to the initiative's market potential.[54]

As explored through the genealogy of refugee and migrant reception in New York and Berlin in chapter 3, there has been a convergence of the hostile immigration environment and Silicon Valley logics. Since September 11, 2001, European narratives undergirding racial discrimination and differentiation have been "overwhelmingly cultural" and "flaunt ethno-racial categories decided on the basis of religious identity ('Muslims' being grouped as a de facto race), national or geopolitical origins ('Middle Easterners'), or members in a linguistic community (Arabic-speakers standing in for Arabs)."[55] This discrimination and differentiation is particularly pronounced in transient spaces such as airports, across news media outlets, and consistently in the context of refuge.[56] Against the backdrop of this framing of the question of the nonassimilability of the "European Muslim" in particular, technologies that single out particular populations through digital refugeeness reinforce these hegemonic discriminatory narratives. Whiteness, as this chapter has shown, flattens refugees into a distinctly racialized homogenous whole, which forecloses agency and self-determination. In the abstraction of the digital, online and offline iterations of refugeeness come together to solidify the purportedly known nature of the proximate other.[57] This has given way to a modus operandi of value generation that is chiefly concerned with reinforcing and selling German generosity through products built on the plight of refugees. Meanwhile, the value of these tools remains essentially negligible for newcomer communities.

Interrogation of refugeeness reveals how technology actors in the digital age engage in categorization that in addition to commanding capital, keeps communities on the move locked in to particular racialized identities predicated on their precarity.[58] By emphasizing and artificially constructing different channels for access, these technologies further entrench these processes of stigmatization of place, people, and practices. This instrumentalization of newcomer subjectivities for technology production provides both the appearance of controlling refugee destinies and, for technologists, a sandbox in which experimentalism on newcomer communities can take place. As these initiatives continue to demonstrate a dearth of understanding of the newcomer's predicament, affected communities are increasingly engaged in refusal.

New York and Berlin are illustrative of profound transformations in cities of refuge. These changes have tended to be twofold: first, the erection of a digital urban infrastructure that bifurcates the city beyond post-Fordist terms; and second, the production of value contingent on the image (but simultaneous absence) of marginalized subjects. The ultimate consequence of this is the technology-driven disciplining of physical movement.

Meanwhile, configurations such as those in Berlin create the conditions under which the market is reinforced, while state actors engage only in a funding and marketing capacity. This allows corporate entities behind refugee tech to keep capital in circulation in the name of refugees, while state actors retain a symbolic veil of proactiveness against the backdrop of the refugee crisis. In contrast to New York, digital refugeeness in Berlin exploits the subjectivities, rather than the physical movement, of moving bodies. In both scenarios, technology actors benefit from extended powers, whether by direct, centralized means or in indirect, decentralized ways. These are the logics that describe how the digital periphery weaves together seemingly disparate geographic contexts in the pursuit of racial capital.

PART THREE

MACHINE-BREAKING, NEO-LUDDISM, AND FUGITIVITY

6

DISCIPLINING MOBILITIES IN THE DIGITAL PERIPHERY

THE DIGITAL PERIPHERY turns borders into "increasingly . . . mobile, portable, omnipresent and ubiquitous realities."[1] When situated within an understanding of the dynamics of racial capitalism, the role of technology deployments *as* the border starts to crystallize. As critical race and migration scholars such as Robyn Maynard[2] and Sandro Mezzadra[3] have argued, the border—whether in its contemporary iteration, particularly across the Mediterranean as a reemergent locus, or through its historical iterations, such as under conditions of Black fugitivity in the Americas   is a site of death and capture. The rapid digitalization of the migrant crisis, and the borderization of technology deployments, expose a colonial continuity of the racializing objectives of the border; it promises "death, removal, and containment" at, between, and beyond the border.[4] Notably, in cities of refuge, the borders are creeping closer to those that seek fugitivity from them, in the form of technological systems of surveillance, control, and exploitation.

In the last few chapters, I have deliberately distanced myself from attempts at making any serious claims about the inner technical workings of the so-called black box.[5] Instead, I have sought to locate how digital urban infrastructures operate on marginalized subjects in particular. This was achieved experientially, through the active witnessing of how these systems shaped everyday experiences of life in digital cities of refuge. Ethnographic methods helped not only to show "how collectives of human and non-human actors emerge, solidify, and evolve over time,"[6] but also to demonstrate how human actors unveil the disposition of nonhuman infrastructure power.[7] In addition to "build[ing] the technologies, implement[ing] them, and us[ing] them in their daily lives,"[8] humans—even in their nonuse or even refusal—confer meaning and reveal partial truths about the technologies. This was inspired in no small part by immigrants' rights and activist endeavors, such as Mijente's #NoTechForICE campaign, which have sought to challenge the technical infrastructures underpinning ICE's deportation machinery, on the premise of their experienced effect within communities.

As laid forth in chapter 2, we can understand the digital periphery through three prisms, namely techno-development, techno-space, and techno-government; these are three distinct vignettes of borderization that we can now apply to New York and Berlin.[9] The first part of this chapter outlines the enduring ways the digital periphery challenges conventional analyses of the world system, via these prisms. Here, I argue for a radical reorientation of scholarship to include the digital periphery as an analytical lens to holistically capture how technology actors reintroduce, reanimate, and reinforce categorization and containment as modes of technology-driven subjugation under contemporary racial capitalism. In the second part, I discuss the implications of the digital periphery for life in urban refuge. I reflect in particular on the implications of the transformation of the post-Fordist city into digital urban infrastructures, skepticism about datafication practices and the possibility for trust, and the emergence of urban migration control and the possibility for sanctuary. Finally, I explore the potential for resistance, revisiting how a handful of initiatives I came across engaged in what I have referred to as neo-Luddite refusal.

TECHNO-DEVELOPMENT

As described by Anita from Reset, service providers such as social and caseworkers have no time or resources to extensively interrogate interventions offered ostensibly free of charge. They're in a position where they're "having to behave like poor people and accept bad deals with high interest."[10] The deployment of modernization tropes and logics by technology actors legitimates their interventions in the lives of undeserving others, in the name of whom experimental digital interventions are iterated. These political groupings are hence either contained in space and subjected to further technological iteration (as was the case with public Wi-Fi infrastructure and smart ID initiatives in New York), or virtually contained through the technical intervention itself (as evident through Integreat, Jobs4Refugees, Taqanu, and many other initiatives coming out of organizations such as Techfugees and the "refugee tech" moment in Berlin). The intervention enables the tech actor to extract (or artificially generate) the raw materials needed to develop further products, and it drives financial capital from philanthropic, venture capitalist, and humanitarian actors as well as governments to the tech actor, which intervenes on their behalf. The digital periphery matters in terms of techno-development, evident especially in how it carries forward existing colonialities of power, through control of subjectivity and knowledge.[11] The treatment of displaced populations in New York City and Berlin are just two examples of this.

The insistence on the "worldview"—to paraphrase RethinkLinkNYC—that frames technology deployments as meeting a socially justifiable need is laced with the same colonialities of power inherited from modernization logics—whether through the imposition of LinkNYC kiosks, the introduction of RFID chips in IDNYC, the digitalization of the inequitable Housing Connect service, or the investment in refugee-specific apps in Berlin. As Safa'a AbuJarour and colleagues demonstrate in their analysis of smartphone usage by refugees, more often than not, refugees rely on similar mainstream apps for obtaining information as the next person.[12] Mirca Madianou as well as Pitso Tsibolane and Irwin Brown[13] have written extensively about technologies deployed in humanitarian and development contexts (often referred to as information communication technologies for development or ICT4D) as colonial practice,[14] with the language of efficiency masking underlying colonialities of power.[15] When Anita invokes a "poor countries" metaphor, she is also pointing to the existence of the same problematic colonial dependencies of ICT4D but embedded in an urban "Global North" context.[16]

More often than not, the technology deployments fail to deliver on the promised distributional efficiencies. Yunus's outrage at IDNYC is emblematic of this. That the promotional selling point of IDNYC tended to focus on free or discounted access to museums and aquariums, all the while making it easier to detain and deport undocumented immigrants, is a particularly disturbing paternalistic and disciplining orientation to incentive structures. It unveils the continued fallacy of modernization theory, demonstrating that so-called *users* of technologies deployed under such auspices are *not*, in fact, the beneficiaries. This is also clear from the deficit in local knowledge input and production, which is treated as an obstacle at best and a hindrance to tech development at worst.[17] Similarly, in Berlin, Hakan remained baffled in the face of the ever patronizing and Eurocentric insistence on discerning refugee usage of smartphones—the mere ownership of a handset by a refugee remained, as previously discussed, an object of great surprise and controversy in the public discourse across European nations. It follows, then, that the refugee is written off as "underdeveloped."

The digital periphery relies on and augments these inherently patronizing, racialized, and neocolonial attitudes toward displaced subjects.

The techno-capitalist orientation of Techfugees toward refugee integration as in need of being offered "as a service" in Berlin is a key moment in which this comes out. For Matteo and Techfugees, techno-development is apolitical; for them, politics must be bypassed in the name of progress, which Matteo equates with municipalities and venture capitalists funding tech initiatives to offer integration. The unchecked imperative to engage with problematic technology actors in a bid to perform "techno-development" in communities of "'know-nots' . . . in need of Western structures and infrastructure," however, reaffirms the enduring modernist nature of development practice.[18] The implicit consensus between the Department of State, New York, and Silicon Valley giants to develop the city's urban infrastructure while also priming it for disaggregated migration control can be seen as one manifestation of Matteo's vision, albeit on a different scale. It also demonstrates that public–private entanglements of this kind are never apolitical. Certainly, in the context of migration governance, these partnerships stand to augment the data-capture capabilities at the center of the surveillance-to-deportation pipeline in New York.

Against the backdrop of the camp voyeurism documented by a sympathetic informant from Techfugees, who incisively noted that "people just want to get to the camps to find needs for their solution," technologists have managed to find urban test-beds malleable to their fantasies about refugee needs.[19] This is inherited from techno-developmental logics, which draw from a long tradition of mediated representation of the needy subject in humanitarianism and development. This is part and parcel of a crucial component of Quijano's colonialities of power matrix, namely control of subjectivity and knowledge. In particular, the underlying messaging of humanitarian imagery is often ethnocentric "moral rhetoric masquerading as visual evidence."[20] As Sarina's experience in Berlin reminds us, refugee tech is nothing special without a "victimized narrative." Sonya de Laat and Valérie Gorin outline a standardized "humanitarian arc" that tends to underpin these reductive forms of representation: (1) "the victim, invariably described as 'innocent,'" in this case, the refugee; (2) "a villain" (which can be a disease, disaster, etc.), in this case, displacement; (3) "a hero, in most cases either a technology or a person of light skin and of socio-economic privilege." In this particular arc, refugee tech is a stand-in for hero-cum-whiteness, constructing and selling the proximate other. This is a continuity of relations of power central to racial capitalism. Indeed, as discussed in chapter 1, racial capitalism has mandated an accentuation of difference along "regional, subcultural, and dialectal" lines, forging these into "racial formations."[21]

These formations continue to have salience, further fortifying the digital periphery through techno-development. As these technological initiatives are further diffused, as per the media theorist Seb Franklin, "a far wider set of socioeconomic logics and practices undergirding the characteristic impositions of the current stage of global capitalism" fester.[22] These practices in turn manifest particular ways of seeing, developing, using, and applying science, technology, and knowledge in society at large, a recurring cycle that perpetuates the production of technologies with accompanying social structures, and vice versa. This gradually expands hegemonic social and economic relations between the digital periphery and those who command racial capital.

Today, well-known technology giants like Cubic (involved in defense contracting with Israel), Intersection/Alphabet (with its track record of involvement in the development of tools used for surveillance), Amazon (with its extensive history of workers' rights abuses and provision of AWS server space that powers ICE's digital deportation infrastructure), and Microsoft (which designed the Domain Awareness System in New York and provides Azure Cloud Infrastructure services to ICE) are embedded within this same system of racial capitalism. They are well-documented enablers of violent policies in different domains that affect migrant populations in particular. As actors that either directly supply both border control and smart city technologies (New York and ICE), while directly or indirectly funding or setting up initiatives focused on refugees and newcomers (in Berlin and UNHCR-supported refugee camps), they engage in a complex and somewhat dialectical exercise of control, relying on both victimization as well as demonization and securitization for legitimation. The digital periphery comes to fruition in part through this interfacing of technology corporations with modernization logics and marginalized populations such as refugees, undocumented immigrants, and communities of color.

TECHNO-SPACE

While newcomers are nowhere near the idealized refugee camp, the metaphor of the camp still elicits an affective reaction among technology companies and policymakers. The digital periphery perpetuates camp-like conditions in its treatment of communities on the move, providing digital terrain on which products can be iterated.

Myria Georgiou defines "digital infrastructures" as two-dimensional sociotechnical systems with "a functional dimension (access, connectivity, use of technologies) and a performative dimension (engagement with technology for seeing and representing oneself and others and enacting citizenship digitally)."[23] These digital infrastructures have attained value in part through the "outernet," which can be understood as the material and planetary movements informing "flows of labour, culture, and power."[24] This comes through in how initiatives such as Jobs4Refugees and Refugee.info generate value from refugeeness alone, absent the refugee. The technology initiative generates racial capital by selling the cultural and political significance of the "vulnerable" refugee, as a potentially lucrative site of extraction, through their apps. Drawing on Lefebvre's "representational space,"[25] digital urban infrastructures manifest digital tropes and sociotechnical imaginaries about migrants in material form. They layer physical infrastructure with symbolic space, which in turn has consequences for the reordering of both. The digital periphery operates in part through this—what I refer to as techno-space, where marginalized and in particular formerly colonized populations are contained physically and symbolically. Through techno-space, the surface area available for the extraction of racial capital is expanded.

Back in New York, towering LinkNYC kiosks on Church Avenue let perhaps the largest concentration of undocumented communities in the borough know that they are under observation, aiding Intersection's linguistic and cultural targeting of "communities of interest"[26] and conscripting spaces of refuge into a surveillant regime. New York has seen digital infrastructures undergirding the physical urban environment, such as LinkNYC, IDNYC and others, retrofitted into an experimental migration-control apparatus (the urban milieu is in other words converted into techno-space altogether). They give rise to what can best be described as "information panics," or what privacy scholars have referred to "chilling effects."[27] Alhasan, the organizer with the Darfur People's Association of New York, along with many other informants from immigrant communities, underscored how it would be inconceivable to connect to public Wi-Fi due to surveillance concerns. Ruha Benjamin's example of hostile architecture—"oddly shaped and artistic-looking bench[es] that make it uncomfortable but not impossible to sit for very long"—is instructive for this.[28] Information panics are representative of human reactions to a hostile digital architecture, which does not strictly render the tool in question unusable or the city uninhabitable, but poses significant hazards and discomforts, making resistance harder.

Similarly, privacy activists with Rethink emphasized that the gaze of surveillance in New York had material and traumatic consequences for marginalized populations, yet there is reluctance to challenge it "out of fear that you might be targeted, or that your community might."[29] For New Yorkers of means and citizenship, interventions such as LinkNYC have become normalized parts of the urban landscape.[30] For most informants in precarious conditions actively looking for ways to organize for everyday survival, refusal of techno-space and an increased engagement in trusted diaspora channels took center stage in fighting what was very much seen as an extension of ICE's security apparatus.

Meanwhile, newcomers in Berlin have been entirely sidelined owing to an instrumentalization of their selves. Here, abstracted subjectivities—that is, *digital refugeeness*—have been transformed into sandboxed spaces for experimentation. Digital initiatives, insofar that they hold the power to classify groupings in the public imaginary, contribute to the objectification of human subjects into manageable others. In so doing, refugeeness attains a technologically mediated racialized disposition.[31] Notably, in Berlin this separates the newcomer from the illusory homogenous whole of citizens while centering German *Willkommenskultur* and generosity. This is how Tom is able to attract funding to Jobs4Refugees. The "outernet" dictates that the imagery and symbolism of the refugee has value (not least as it continues to serve as a sensationalist spectacle in the tug of war between progressive and, conversely, nativist and xenophobic politics), while refugee tech, in turn, has material consequences. The symbolic or virtual space demarcated by Jobs4Refugees contains twenty-two thousand digital refugees (or avatars) within it. The actual number, in reality, does not matter, the digital representation of subjectivities does—or rather, the symbolic value ascribed to Black and Brown "likes" and images. Remember that for tech entrepreneurs like Tom, measuring the extent to which refugees actually used their initiative wasn't necessarily a "highly effective measure to evaluate."[32]

The historical echo of these digital forms of enclosure harks back to the early use of human collateral to "raise a significant amount of cash and credit" in the colonial South of the United States through mortgaging slaves.[33] The digital periphery, however, carries out the entrapment computationally, and often out of sight, increasingly abstracting the extractive and violent relations of power to the point of perceived intangibility. Both cities invariably constitute what Nicholas Mirzoeff refers to as "white space," where regime "oversight," by analogy of the plantation overseer, polices and "ensures maximum production and minimum resistance"[34] by removing the possibility for agency altogether.

Today, as evidenced in New York and Berlin, the digital periphery works through techno-space to expand the surface area available for technology deployments. In so doing, it extends our traditional conceptualizations of the border to encompass the boundaries of otherness and encodes the latter in technologically exploitable terms. The allure of technology in both cities for solving complex problems related to displaced bodies—or as is mostly the case in my examples, find problems for deterministic solutions—relies on the willful ignorance of the experiences of violence and precarity along racial lines, in service of the continued technology deployment.

TECHNO-GOVERNMENT

In New York and Berlin, governments and institutions rely on technology actors and their digital urban interventions, symbiotically, to identify, exploit, and control immigrant communities. Technology actors share in migration governance by deploying interventions that instrumentalize urban migrant environments or subjectivities, thereby constraining movement and exploiting fugitivity for racial capital. These cities of refuge demonstrate how the digital periphery works along techno-governmental lines.

Keller Easterling has shown how this is possible because urban space is often a site of "multiple, overlapping, or *nested* forms of sovereignty, where domestic and transnational jurisdictions collide."[35] This sustains technological experimentalism, data extraction, and the mobilization of vast amounts of capital for profit and drives symbolic and political capital from governments and local authorities to tech actor; what Easterling would call the exercise of "extrastatecraft."[36] Indeed, governments can erect a mirage of order and control through tech deployments while allowing technology corporations to become an essential part of the performance of governance, even if they are not openly acknowledged as such. Liberties are granted to technology corporations in exchange for enabling and sharing in governance historically at the expense of migrant laborers, prohibiting, strikes, lowering minimum wages, exploiting their data, and surveilling them.

The digitalization of cities has implications for the distribution of power between city and technology actors. In the context of New York City, there is a greater centralization of control and surveillance powers with authorities, including at the federal level. This is reinforced through new technological innovations, which in turn inserts technology actors in the business of governance. As chapter 4 showed, this comes through in the deployment of Intersection's LinkNYC kiosks. Mayor Bill de Blasio not only situated the city as a "testbed for new technologies [transforming] the relationship between city government, community, and the tech industry" in October 2017,[37] but went as far as to appoint Facebook, Google, IBM, Microsoft, and Verizon to its NYCx Advisory Board and several other elite city committees.[38] When the partnership between the Mayor's Office of Immigrant Affairs and LinkNYC was announced, purportedly with the objective of providing information to immigration communities, it was already a known fact that ICE had conducted significant deportation raids across the city, and that its work was supported by software infrastructure provided by the very tech giants who had just been given unfettered access to New York's digital real estate.

While this techno-deterministic orientation to governance at the macro level exposes a revolving door between the tech industry and government, it is also indicative of design practices that transpose nested sovereignties to the street level.[39] For instance, my conversations with Intersection demonstrated that middle managers working at the level of deployment were genuinely convinced they were providing a public good. I recall how they explained that public consultations in the design phase of the kiosks had adequately involved the community. The perception among communities—who are neither a homogenous whole nor always in agreement—was, needless to say, a far cry from this understanding. My informants insisted that surveillance capabilities had *not* been presented to them at the design consultation stage. These capabilities included cameras, Bluetooth sensors, and data collection defaults of several data points that could be used to establish demographics, such as device type, phone language, a hash key for Wi-Fi reconnection, and email address.

Similarly, the proposed RFID upgrade to IDNYC was a later addition that the city incessantly pushed, despite vehement resistance from activists and public defenders on grounds of potential function creep and data exploitation by ICE. As critical design scholar Sasha Constanza-Chock notes, "Too often, design teams only include 'diverse' user personas at the beginning of their process, to inform ideation."[40] These design decisions ultimately serve to give impetus to urban governmental entanglements between the state and the tech industry, which provides the sensors through which data on communities is collected, processed and shared with the state, and used in the constant reiteration of new product lines.

Conversely, in Berlin, technology actors have been left somewhat to their own devices in the context of technology interventions for refugees in particular—they step in as outsourced providers of integration. By cutting through the red tape associated with the German bureaucracy, and with authorities remaining either ambivalent or actively opting in to the intervention, technology actors occupy not only significant agenda-setting powers but also positions of authority concerning refugee integration. As I explored in chapter 5, at the macro level, state actors such as BAMF were getting involved only as potential funders. I also learned that larger companies, such as Salesforce, Google, and Facebook, were investing in the research and development of refugee tech. This laissez-faire approach enabled experimentalism with vulnerable populations and rendered oversight difficult. The refugee tech initiatives I examined in Berlin did not have a public consultation process by which they were legitimated. In the case of Integreat, I learned from the representatives—as was the case for Jobs4Refugees and several other initiatives—that the only usability testing they conducted for their tools were with their clientele of municipalities. They relied on the "word of mouth," of municipal workers as proxy data for refugee experiences, as one of my informants claimed.[41] It is curious that a tool designed to deliver crucial information to refugees did not have a feedback loop to assess the delivery of this information. Constanza-Chok's description of design-driven exclusion best summarizes the implication of the approaches taken by urban refugee tech entrepreneurs: "If you're not at the table, you're on the menu."[42] In both cities, technology actors benefit from extended powers granted through the situatedness of technology giants as essential to governance.

As discussed in chapter 1, racialized subjects were conditioned to fit the servile periphery in service of Europe's source of value in trade during the early formation of racial capitalism. Secondly, as a source of further value extraction to meet the production demands of the world market,[43] the proximity of newcomers and immigrant populations to Western urban centers today has generated a greater need for creating distinct strata in which these proximate others can be contained (hence refugeeness). Production, and in particular, tech production, can be justified under the auspices of these strata and on the backs of these communities. This reinforces the colonial aspect of technology production, maintaining the flow of racial capital among the tech industry, funders, and government.

A final feature of the techno-governmental vignette is limited to cities with more extensive smart infrastructures, such as New York, where there is a stigmatization of technology refusal. Instances of anti-LinkNYC activism, through breaking cameras on kiosks as per the case of Juan Rodriguez, have been highlighted as vandalism and criminal activity. A more recent development includes the activism against the new contactless card OMNY system that has started to be deployed across the Metropolitan Transport Authority (discussed in chapter 4).

The digital periphery justifies and encourages techno-governmental relations of control through the mandates of these interventions as service providers and public goods, and their intersection with policing, via stigmatization, criminalization, and data sharing.

LIFE IN DIGITAL URBAN REFUGE

Life in urban refuge, as experienced by occupants of the digital periphery across New York and Berlin, is a stark reminder of how the city as a destination of sanctuary has been transformed in two major ways over the last thirty years: first, through the erection of digital urban infrastructure that bifurcates the city beyond post-Fordist terms; and second, through the production of speculative value contingent on the image (but simultaneous absence) of marginalized subjects.

Under post-Fordism, the formerly demarcated "ghetto" is excluded, rather than in an active state of direct exploitation.[44] However, the combination of these post-Fordist changes, along with the emergence of the digital periphery through digital urban infrastructures, make possible the simultaneous exclusion, domination, and exploitation of undesirable populations. Information panics are one example of this. Technologies such as LinkNYC evoke crises, which accelerate the digital sedimentation of the post-Fordist sociospatial order through regulating how subjects can move in urban space, based on characteristics such as immigration status and race in particular. In the "mobile outcast ghetto," of which undocumented and other vulnerable migrant populations can be designated a constituent part, "logistics orchestrates the control and management of surplus populations, keeping them in their (social and economic) place, even as they move about the city."[45] The insertion of public Wi-Fi kiosks and its correlation with deportation raids demonstrate how limited vulnerable migrant populations are in terms of physical movement lest they be sensed by adversarial urban infrastructures should they venture into the digital "citadel."[46] Bruce Sterling's speculative description of how the smart city automates segregation and the generation of a "mobile outcast ghetto" is incisive in this regard.[47]

The digital periphery creates the conditions under which the simultaneous expulsion and inclusion of undesirables are mutually reinforcing phenomena that allow for the generation of racial capital. This is the second transformation of cities as refuge: the digital periphery is as much about the categorization and containment of marginal physical bodies as it is about the exploitation of their digital avatars. Beyond the bifurcation of the city into mobile "outcast ghetto" and digital "citadel," digital urban infrastructures use the image of those in the digital periphery to legitimate their intervention. For instance, Intersection uses the ad screens on its kiosks to advertise them as access points to immigration services while also broadcasting occasional "fun facts" about immigrant histories for passersby.

The municipal ID system IDNYC also works to advance both of these transformations. Without completely replacing existing IDs, they stratify city populations into those at greater risk of being stopped and potentially detained on the suspicion of being undocumented, versus those with regular IDs (once again invoking the above-mentioned bifurcation). These interventions single out particular populations, thus reinforcing underlying discriminatory narratives. In Berlin, several app initiatives generate artificially distinct channels for access, entrenching ethnic and national differentials—not unlike the "mobile" ghetto. As AbuJarour has shown in her work, refugees and newcomers hardly use these distinct tools; nevertheless, their discursive power is not to be underestimated. In cities of refuge like New York and Berlin, technology often acts as a gateway for authorities and institutions to appeal to narratives of gratitude and victimhood. This is particularly problematic against the backdrop of a hostile global immigration environment, in which the politics of the welfare state are inextricably tied up with the xenophobic posturing, by right-wing parties in particular, of racialized populations as abusers of public benefits.

Across both New York and Berlin, interventions made in the name of displaced populations demonstrate a continued neoliberal move toward conjecture as the engine of value creation. In a limited sense, applicable almost exclusively to communities living under conditions of suspended or limited rights—for instance, by being perceived as less deserving of a right to work—it emerges that the "direct labour of humans is no longer of exchange or use value in digital capitalism."[48]

Both case studies have demonstrated that digital urban infrastructures work antithetically to their purported objective (often framed in terms of access, connectivity, and welfare). While structural inequities are frequently invisibilized under digitalization efforts, they endure. Equally, theoretical access to the services through extending access to service "interfaces" (e.g., the ability to access the online application for affordable housing, as is the case with NYC Housing Connect) does not mitigate the problem of gatekeeping, and therefore does not guarantee procedural fairness and better outcomes. The tendency has instead been to augment and exacerbate existing discriminatory practices already faced by communities living under conditions of urban refuge.

TRUST, TECH SKEPTICISM, AND SURROGATE DATA

Shoshana Zuboff has referred to the turn in the usage of metadata as the function of a new era of surveillance capitalism, which "lays claim to private experience for translation into fungible commodities that are rapidly swept into the exhilarating life of the market."[49] This, she claims, is a fundamental shift in the economic model of capitalism. Under surveillance capitalism, "behavioral data that were once discarded or ignored were rediscovered . . . as a means of generating revenue and ultimately turning investment into revenue. [Users] became a means to profit in new behavioral futures markets in which users are neither buyers nor sellers nor product."[50] Zuboff presents a model under which users provide the "raw materials" for capitalism through their digital behavior.

In this book I have distanced myself from this particular orientation for reasons we captured in chapter 2, namely that (1) surveillance, through categorization and containment, has always been a driving force of racial capitalism (a concept with which Zuboff scarcely engages) and (2) my encounters in New York and Berlin demonstrate that technology deployments do not, in fact, depend on meta- or -behavioral data. Rather, they depend on what can be thought of as surrogate data—that is, data that is not generated by those who they purport to pertain to. Instead, this data is based on conjecture *about* abstract racialized categorizations of identity, forged out of white imaginaries about the other.[51] This can be considered "artificial" intelligence, insofar as the data is abstracted, speculative, and in many ways representative of fictitious approximations.[52] Take away the human futures markets and the liberal usage of behavioral data, and racial capitalism still secures capital through surrogate data extracted from the digital periphery Rather than behavioral data or human futures markets, this is given shape by (1) panoptic aesthetics embedded in the urban environment, actively imparting insecurity on undocumented migrant populations, who in turn resist them; (2) obfuscation of the actual extent of control, which forecloses contestation of the ways in which digital urban infrastructures do (or do not) describe, categorize, and lay claim to migrants; and (3) conjecture about migrant populations.

First, perceptions of data collection matter for how individuals experience marginality. For communities contained in the digital periphery, skepticism about digital infrastructures is tantamount to skepticism about institutions and the governing class. In privacy discourse, critiques of such practices are often framed in terms of what Evgeny Morozov calls "data extractivism," a logic that sees users as "valuable stocks of data" for whom

> technology companies, in turn, design clever ways to have us part with that data—or at least share it with them. They need this data either to fuel their advertising-heavy business models—more and better data yields higher advertising earnings per user—or they need it in order to develop advanced modes of artificial intelligence centered around the principle of "deep learning."[53]

However, in digital cities of refuge, skepticism about access to data goes beyond this technical analysis. Chapters 4 and 5 illustrated why controversies around the use of meta- and behavioral data have become particularly rife in recent years. The presence of always-on technologies with inexplicable (and often unreported) sensors, in particular, pose a significant data-extraction risk to migrant communities, whose lived everyday realities are shaped by intensifying ICE deportation raids and unsolicited reporting. The prospect of the RFID-enabled IDNYC, allowing for more detailed tracking of locations and demographics, even if only in theory, is thus resisted. This is hardly surprising given the association of the existing IDNYC system with several immigrant detentions by sheer virtue of being presented to law enforcement. These tools, then, while panoptic—irrespective of actual surveillance capabilities—impart insecurity on migrant populations, whose resistance does not necessarily dissuade the continued insistence on the deployment of the technologies.

Second, and even more perniciously, my informants indicated a general lack of clarity around who has access—especially at the federal level—to data coming through digital urban infrastructures including Wi-Fi kiosks, automated number plate reader data, OMNY cards, and more. These fears are not unfounded. John Cheney-Lippold describes in great detail how a number of discreet categorizations of behaviors, based on data *about* how people behave on the internet, are classified into many different algorithmic brackets (e.g., "male," "Brown," "programmer," "artist")—what he calls "measurable types."[54] Ruha Benjamin explains how the New Jim Code of Silicon Valley uses this model of algorithmic classification in tandem with carceral logics inherited from Jim Crow to, among other things, describe hundreds—perhaps even thousands—of racial classifications, each with an accompanying observed algorithmic behavior.[55] It stands to reason that these interventions likely operate according to similar logics, at the technical level—especially in hostile immigration environments, where organizations like Mijente and Brooklyn Defender Services remind us that the availability of a vast number of data points can be instrumentalized by federal immigration enforcement officers for location-tracking, detention, and eventual deportation. However, the general obscurity and inaccessibility of how these systems operate, and who has access to them, makes it nigh impossible to challenge the ways they do (or do not) describe, categorize, and lay claim to migrant identities.

Finally, as evident in Berlin, there is a discrepancy between the platforms being built by technologists (often with no experience of displacement and no history of engagement with displaced communities) versus those used by refugees in everyday information architectures. This knowledge should be sufficient to reconsider whether these interventions provide anything of value at all. However, tools like Facebook pages provide new avenues for conjecture, which combined with reductive key performance indicators allow refugee-tech initiatives to falsely equate their Facebook audience of "likes" with "refugees." This is what I have referred to as digital refugeeness.

URBAN MIGRATION CONTROL

Engagements with technologists and affected communities in both cities unveiled how technology companies work through modernization logics, space, and governmental entanglement to categorize and contain displaced subjects. The variation across each city is emblematic of how the modes of subjugation central to racial capitalism, categorization, and containment continue to play a role today, with particularly egregious consequence for displaced communities. The digital periphery has, in other words, given further impetus to the disaggregation of migration control, bringing the border closer to both the city and the body. This has implications for how technological deployments in contexts of urban refuge, and migration control broadly, should be viewed going forward.

As digital urban infrastructures increasingly govern how people access services, connect, communicate with authority, work, access housing, and utilize immigration services, it is important to analyze these interventions through the lens of the digital periphery, showing how technology deployments such as those covered in this book are in fact about mobility governance and contemporary practices of bordering. Understanding how digital urban infrastructures specifically exercise control as borders allows future research to expand the scope of inquiry, when investigating **both** borders and border technologies, widening and updating both fields of research. Below, I delineate the two forms of informational control at the center of these technological deployments.

The technologies reviewed in chapters 4 and 5 are engaged in predominantly two forms of informational control, which have implications for how individuals epistemically—and thus also physically—move through refuge, resettlement, and integration. One form of informational control is the ability to force the voluntary or involuntary surrender of information from an individual to the authority requesting it, whether directly or indirectly.[56] This occurs through biometric technologies used for registration, digital vouchers, and surveillance, as well as tools used to collect personal data, metadata, or construct surrogate data in more subtle ways, such as monitoring browsing behavior, device characteristics, transactional histories, telecom data on call and messaging habits, and image scraping. I class these as information control technologies (ICoTs) for solicitation. The second type of informational control is the capacity to control what information people have to act on, precipitating certain action or inaction, deemed desirable or undesirable by the particular authority.[57] This category includes information-dissemination initiatives, such as tools that provide information to help people integrate into a new environment, as well as tools that help them navigate bureaucratic processes—what we might class as ICoTs for dissemination.

Both types of informational control converge at times, particularly in technology-driven migration governance. It is a common configuration that limited information or access is provided, often in exchange for extraordinary amounts of data. Cities are also sites in which this convergence plays out. In New York, refugees, asylees, and undocumented immigrants alike are given access to digital services purportedly intended to improve life in urban refuge, at the expense of giving up data that could be used to detain and deport them. This gives rise to information panics and system aversion, further compounding precarity, insecurity, and, by extension, containment. In Berlin, newcomer communities are asked to potentially engage with dozens of apps, proven to be more or less void of functional value, specifically designed for refugees. Their epistemologies are sidelined in favor of an artificial one that drives racial capital. In this way, start-up capital is kept in circular motion among funders, the state, and technology initiatives.

As populations who do not possess the same level of protection as citizens, occupants of the digital periphery—that is, communities on the move and other racialized displaced communities at large—are used as experimental subjects. They are subjected to a technology-mediated hostile immigration environment, inherited from a century of anti-immigrant policies across both cities, as explored in chapter 3. While technology deployments in migration contexts are framed in emancipatory language, they can also be described as digital Trojan horses for depoliticized but austere xenophobic policies. Within the digital periphery, racialized noncitizens are exploited for racial capital.

The ways in which urban migrant environments and subjectivities are datafied and commodified reveal the inner workings of the digital periphery. With cities posturing as sanctuaries while facilitating the rapid and lucrative weaving of Silicon Valley into the fabric of urban and migration governance, they have become a legitimizing ground for tech companies, a final frontier for the exploitation of those fleeing persecution and hunger. This has profound implications for the future of resistance, fugitivity and sanctuary, especially amid the more recent rampant deployment of digital urban infrastructures with artificial intelligence capabilities, on the back of the AI hype that has dominated mainstream discourse since the launch of ChatGPT in November 2022.[58] Digital urban infrastructures are segregating and bordering cities into vastly different forms and configurations of privileges. This transmutation, staying with the current trajectory, stands to intensify in years to come.

MACHINE-BREAKING AND NEO-LUDDISM

I have sought to sketch out how the digital periphery is an enclosure embedded within the continuing legacy of racial capitalism. Throughout chapters 4 and 5, however, my informants also described routinized practices of refusal. The fear of deportations against the backdrop of an advancing digital urban infrastructure had exacerbated the information panic in New York City, leading communities to engage in such practices of refusal as urban concealment. Sousveillance and neo-

Luddite tactics were and still are on the rise among immigration, privacy, and antipolicing activists in particular, through organizations such as RethinkLinkNYC, the Immigrant Defense Project's ICEwatch, and the mobilization of alternative information channels through the e-diaspora.[59] RethinkLinkNYC, which organizes with undocumented communities, is fighting the kiosks in particular because the technology stands to obstruct future resistance. To Rethink activists, the dynamics of speaking truth to power are stifled in similar ways for immigrants as for survivors of domestic abuse who are expected to report their abuser. Knowing that one is being watched in the first place is experienced as a looming threat. This is why Rethink supported Juan Rodriguez when he was apprehended under the auspices of mental instability. It is also why it plasters LinkNYC kiosks with warning stickers, printed with the motif of a yellow warning sign, cautioning passersby that a camera is, in fact, present on these mundane boxes they walk past every day.

Across both cities, the diaspora was resisting and organizing in groups designed with their communities in mind. Refusal in these groups was not simply a rejection of technology that was either unevenly distributed or negatively impacting them. It was about the power structures, logics, and policies that powered the technology in the first place. Sarina was organizing G100 in Berlin in an effort to "reposition refugees as the ones producing the knowledge, as owning their space, and *you* as the guest, where previously we were the guests."[60] For her, G100 was an epistemic reclamation, a negation of refugeeness altogether. Spaces like Syrische Frauen created safe havens, bordering away the German gaze—as well as in-diaspora male-dominated online discourse—and its insistence on a linear, top-down integration policy. The group has flown under the radar because information about it was disseminated *by* members of the diaspora, *for* the diaspora. Technology-augmented spaces for displaced populations are less inundated by the sensationalism afforded to refugee tech, even if "mundane" in form. It is an inherent refusal of the gaze that would otherwise view the space of digital refuge with market potential front of mind.

Through techno-development, -space, and -government, this moment, while certainly shaped by technology, is not strictly about technology. Rather, it is about how people are racialized, how race is instrumentalized, and how bodies and spaces are exploited as a result. As Ruha Benjamin aptly posits, race itself, in this way, is a form of technology: "Racism is, let us not forget, a means to reconcile contradictions. Only a society that extolled 'liberty for all' while holding millions of people in bondage requires such a powerful ideology in order to build a nation amid such a startling contradiction."[61] Remembering that Cedric Robinson teaches us that those fleeing racial regimes are fugitives,[62] the resistance against framing their fugitivity in terms of technological terms matters profoundly.

Other initiatives had attempted to demonstrate an alternative vision for technology, but they too faced funding difficulties, largely due to their refusal to benefit from conditions of suffering. In line with sympathetic immigration lawyers, they have adjusted to the reality of New York as a site for migrant violence. These initiatives have an acute understanding of the fears around deportation raids in particular and actively self-sabotage by taking a data-minimizing approach. They encourage their clients to be mindful about the information they share with them. Good Call NYC was one such initiative that intentionally locked itself out of accessing or granting access to data from their clients, who are often calling as a last resort in attempts to rectify conditions of unjust incarceration. This, too, is a negation of the power structures, logics, and policies reproduced by the dominant paradigm in the tech industry. It is a refusal to accept the "social cost of technological progress" promised by the moment, which is otherwise a prerequisite to the production of wealth in the digital periphery.[63]

Racial capitalism orders and organizes the production (and deployment) of the very machines Luddites originally rallied against. In the nineteenth century, a group of textile workers in Manchester, England, known as Luddites, engaged in acts of civil disobedience in response to the introduction of machinery to their industry. They left letters around the city where they had destroyed stocking frame machines, in protest against the decreasing standard of living and wages.[64] Their activism follows a long tradition of workers struggling against the destructive powers of laissez-faire capitalism in northern Britain, where the transition to greater automation in the name of progress had led to unmitigated social devastation and decline for the working class. For the Luddites, destruction was a method of last resort to preserve what little remained of a "world . . . on the verge of destruction."[65] From the twentieth century and into our current moment, social movements "who question the predominant worldview, which preaches that unbridled technology represents progress," are essentially Luddite.[66] They struggle against the same process of wealth accumulation as a function of technological deployments, at the expense and on the back of racialized, working class, and often migrant communities. Taking a page out of the abolitionist and neo-Luddite orientations, then, it is incumbent on us to interrogate and understand exactly how the structures we wish to undo are upheld, obfuscated, and reinforced by violent technologies that promise greater efficiency, convenience, and security—technologies that uphold the digital periphery.

In 1990, Chellis Glendinning published her "Notes toward a Neo-Luddite Manifesto," in which she applauded the Luddites for taking laissez-faire capitalism to task for enabling the "increasing amalgamation of power, resources, and wealth, rationalized by its emphasis on 'progress.'" In the twentieth century and at the time of Glendinning's writing, neo-Luddites were—and largely remain—the "activists, workers, neighbors, social critics, and scholars . . . who question the predominant worldview, which preaches that unbridled technology represents progress." She formulated three principles undergirding neo-Luddism:

1. *Neo-Luddites are not anti-technology.* . . . What we oppose are the *kinds* of technologies that are, at root, destructive of human lives and communities. . . .
2. *All technologies are political.* . . . They tend to be structured for short-term efficiency, ease of production, distribution, marketing, and profit potential—or for war-making. As a result, they tend to create rigid social systems and institutions that people do not understand and cannot change or control. . . .
3. *The personal view of technology is dangerously limited.* The often-heard message "but I couldn't live without my word processor" denies the wider consequences of widespread use of computers (toxic contamination of workers in electronic plants and the solidifying of corporate power through exclusive access to new information in databases).[67]

In a 2014 *Forbes* article, a new wave of emergent Luddism is described as situating its focus squarely on privacy, an alleged diversion from the nineteenth-century revolts against the job-stealing machine.[68] Contemporary invocations of the term, however, have not been in dialogue with critical race and postcolonial scholarship around themes such as refusal, abolitionism, and violence, despite their pertinence to how and why contemporary technology deployments are especially resisted in racially marginalized contexts—the border is often the frontline for a struggle to maintain what remains of dignity and agency. Syed Mustafa Ali's "fugitive decolonial Luddism" describes the nodes that connect racial struggle and fugitivity with Luddism: "Luddism, as an active, oppositional stance toward specific technological developments, is usefully retrieved through 'entanglement' with decolonial computing, and further enhanced by the adoption of a fugitive orientation toward surveillant datafication drives."[69]

Much like Fanon's "triumphant *communiqués*" of missionaries seeking to implant foreign influence "in the core of the colonized people,"[70] techno-solutionism speaks religiously of a scientifically perfectible world, albeit with one caveat: it does not consider itself a source of the imperfect. Fanon's violence, then, as "a cleansing force [that] frees the native from his inferiority complex and from his despair and inaction," must in the digital age include a Luddite imperative: to first and foremost resist, sideline, co-opt, and break the containment chambers and borders written in ones and zeros. The tactics used to navigate refuge in the digital periphery thus far give limited but novel insight into what a decolonial neo-Luddite approach to digital technology interventions in the twenty-first century might look like.

These examples illustrate why epistemologies of migrant survival must be conceptualized as fundamentally opposed to the dominant information regime, and as an abolitionist and fugitive undertaking. While both cities have been a site through which the experimental logics of the digital periphery have been especially revealing, it has also set the stage for the accentuation of modes of techno-racial resistance. Through a reconciliation of practices of refusal documented throughout this chapter, it is apparent that alternative emancipatory decolonial imaginations concerning technology are possible.

CONCLUSION

IN THE AFTERMATH of the Syrian refugee crisis of 2015, the possibility of information communication technologies to ameliorate conditions of displacement was met with great elation. The provision of information, housing, employment, resettlement, and integration were squarely situated within the domain of the technology sector. The Migration Policy Institute hailed the smart city for harboring particular potential for addressing integration, socioeconomic access, and inequality among refugees and migrant communities at large, valorizing "digital tools [for navigating] local services," gig economy jobs, and sharing-economy initiatives.[1] Combined with the concurrent emergence of initiatives like Techfugees, which falsely conflated the plight of refugees with technical problems waiting to be addressed by the "hackathon," this techno-solutionist and techno-chauvinist orientation[2] fundamentally banished considerations around bordering, the adjacency of the technology sector to the security industry (and the military-, prison-, and border-industrial complexes), and potential transformations in relations of power to a mere afterthought. Yet, it was orientations like these that permitted the accelerated intrusion of Silicon Valley into the lives of displaced populations. Against the backdrop of a free/libre/open source software community that was incrementally being co-opted and recruited into mainstream corporate Silicon Valley giants, tech companies like Facebook, Microsoft, and Google adopted the usage of emancipatory language to situate themselves as political stakeholders working toward justice-oriented goals.

In negation of this particular epistemology, and in contribution to more critical epistemologies that are void of considerations around race and coloniality, digital urban infrastructures and their impacts on communities on the move demonstrate how the techno-solutionist and chauvinist orientation, and its ascribed actors, have transformed migrant environments and subjectivities into contested spaces in the battle for racial capital. Within the bounds of the two distinct geographies—New York City and Berlin—we see how race, border, and migrant entanglements have intensified in the digital age, giving rise to the digital periphery through disciplining mobilities and exploiting so-called unruly bodies via techno-development, techno-space, and techno-government. This nevertheless implicates how researchers understand and study race, migration, and technology, which must be examined as interconnected and mutually constitutive phenomena going forward.

Through this particular analytical lens rooted in the Black radical tradition, critical migration studies, and science and technology studies, I have attempted to contribute to critical and emergent debates on migration technologies, as well as critical race and digital studies. My hope is to push critical technology discourse to acknowledge the centrality of migration and the management of mobilities to digital manifestations of racial capitalism, and furthermore, to center the role of racialized technology production in critical border and migration studies.

In these concluding sections, I reflect on the implications of the digital periphery for a future research agenda. Here, I make the case that the digital periphery reveals new and crucial ways of viewing cities, borders, interstitial geographies such as refugee camps and detention centers, and global movement as interconnected nodes in the production and reproduction of racial capitalism. Finally, I pose a series of open questions about the possibility of effective resistance.

CITIES, REVISITED

Life in digital cities of refuge is a stark reminder of the transformations that have occurred in smart urban environments, and the terrains that racial capitalism will avail for value generation. In chapter 4 and 5, I observed these transformations empirically, actively witnessing how migrant community environments (in New York) and subjectivities (in Berlin) were exploited for profit and prestige. These case studies show how increasingly interwoven Silicon Valley is in the fabric of urban governance. I referred to the city as a digital antisanctuary to emphasize how fugitivity is stunted and exploited through an implicit consensus between the city, Silicon Valley, and immigration enforcement. This in turn is sparking neo-Luddite refusal of datafication through urban concealment among affected communities. A larger implication of this is that the digitalization of cities reconfigures existing distributions of power between city and technology actors. In New York, the market reinforces the state by giving it a greater centralization of surveillance power. This is emphasized through new technological innovations, which in turn insert technology actors in the business of governance.

In Berlin, digital refugeeness allows the corporate entities behind refugee tech to keep capital in circulation in the name of refugees, while state actors retain a symbolic veil of proactiveness against the backdrop of the refugee crisis, as explored in chapter 5. While this is undoubtedly exacerbated in the digital era, chapter 3 also traced how early twentieth-century anti-immigration sentiment in Germany seeps into how the state negotiates its relationship to foreign (i.e., non-EU) immigrants. It is politically expedient to advance the myth that refugee preferences for literacy, flexibility, and stability are different; terms such as "resilience" are often used to situate refugees and migrants as having a greater appetite for precarity than "regular" people, justifying their exploitation and abuse. In the name of techno-urban entrepreneurship, spheres under the purview of local authorities and the state have in this way increasingly been encased in privatized digital instances specifically tailored toward refugees.

New York and Berlin are key examples of how the digital periphery weaves together seemingly disparate geographic contexts in the pursuit of racial capital. However, they are by no means the only ones. While I was limited to just two field sites in this research, due to funding and timing constraints, it is paramount that future research engages in mapping the digital periphery across further cities of refuge with rapidly digitalizing infrastructures. For example, Nairobi and Barcelona have in recent years seen major digital transformations that likely have significant implications for how their sizeable populations of refugees move in the city. Under Mayor Ada Colau, Barcelona has led a public effort to purportedly restore citizen control over data, but in so doing has inadvertently articulated an understanding of urban citizenship through which data is encouraged to be actively and voluntarily shared.[3] This model, framed under the auspices of the digital data commons, is equated with good citizenship, and could potentially advance a scenario in which a willingness to engage in more data-sharing is equated with a greater amount of privileges and prestige. This could have dire consequences for refugee and "irregular" migrant communities, who live in fear of the Centros de Internamiento de Extranjeros.[4]

While structural inequities are often hidden under digitalization efforts—even at times co-opted as the problem justifying the deployment of the particular technology—they endure. Ultimately, the analysis in chapter 6 made it apparent that digital urban infrastructures work antithetically to their purported objectives, while reliant on emancipatory framings such as the myth of extending access (to what?), connectivity (on whose terms?), and welfare (at what cost?). The tendency has instead been to augment and exacerbate existing discriminatory practices faced by communities living under conditions of urban refuge. As such, interrogating further digital cities of refuge in techno-developmental, techno-spatial, and techno-governmental terms can help shed light on how the digital periphery is constituted and exploited in different contexts.

INTERSTITIAL GEOGRAPHIES

The digital periphery squarely positions this technological paradigm within a historical moment of racial capitalism and opens up the possibility for a new research agenda more aligned with the output from the burgeoning field of critical race and digital studies. As I sought to expand conventional and distinctly physical conceptions of the border, I also explored how new forms of digital enclosure perform migration control in cities of refugee. Throughout the process of this research, however, it has become clear that there is a need for a more comprehensive research effort focused on mapping the global digital periphery across a number of interconnected geographies. Furthermore, the greater impetus given to the diffusion of sophisticated technologies of control at the time of writing, under the continued pandemic conditions of COVID-19, warrants a transnational effort to analyze the contingency of these deployments on marginalized displaced communities.

As mentioned in chapters 1 and 2, refugee camps and borders have been the sites of a great many research efforts concerned with the deployment of biometric and other emergent technologies in particular. For example, Helle Stenum's work has argued that biometric deployments in camps, in particular, were increasingly carving the border into the body.[5] Compounding this, blockchain-based ID initiatives risk hard coding the border in the body in potentially irreversible ways, rendering what Keren Weitzberg has called "machine-readable refugees."[6] Developments ranging from IrisGuard to ID2020 and World ID / Worldcoin are key examples of this. These practices constrain the agency and choice of communities on the move, their "survival strategies, their hopes for a better future, none of which can be captured on a digital scanner or encoded into a database."[7] However, resulting harms are not simply unforeseeable side effects of systems deployed with the interest of displaced populations in mind. As described in chapters 1 and 2, and as evidenced in chapters 4 and 5, the digital periphery comes into existence in no small part due to the perception of marginalized communities as themselves constitutive of an emergent market. It allows technology actors to engage experimentally not only on terms related to innovations in security (i.e., national security, immigration enforcement, or border control) but also, simultaneously, on humanitarian grounds.

Even along routes of transit, the digital periphery is there to provide. Investments in research and development for experimental border control and policing technologies by the European Commission, under the Horizon 2020 grant,[8] have amounted to €1.7 billion and included systems such as unmanned drone surveillance to aid in Frontex's interception of migrants.[9] The fund also enabled the development of proto-eugenicist "lie detection" systems based on "micro expressions"—a practice by which someone's momentary expressions are taken as an indication of their emotional state. The technique, which is based in controversial pseudo-scientific fields of phrenology and physiognomy, follows similar logics to research enabling the systematic and state-sanctioned oppression, internment, and murder of Uighur Muslim minorities in China.[10] Research published by Duan Xiao-dong and colleagues, for example, "create[d] a face database of ethnic groups and extract[ed] facial features by using face recognition technology" among "Tibetan, Uighur and Zhuang" ethnic groups.[11] The digital periphery, in this way, operates across vast distances, tying together marginal geographies for technology-driven extraction of racial capital.

Working in concurrence with these overtly nefarious interventions, more subtle applications of information control technologies (ICoTs) under humanitarian auspices also remain in place along paths of refuge. These include, but are not limited to, collaborations between humanitarian organizations such as the joint works of Mercy Corps and Google, emitting Wi-Fi across these pathways to promote the usage of their informational tool for refugees, and the relationship between World Food Program and Palantir to name a few. While on fieldwork, I encountered several humanitarian workers who had been contracted by a major humanitarian organization to oversee technology deployments such as these. They lamented that they had been hired to create "things that don't work, creating things for show," noting that a mere "download is a win in the humanitarian sector."[12] Here, a win is equated with meeting funder requirements and securing incoming flows of funding in the future, rather than metrics rooted in the needs and welfare of communities on the move.

A research effort that used the digital periphery as a lens would place these sites in dialogue and integrate other interstitial spaces such as prisons and detention centers. If the conventional lines drawn around borders have moved, so research must seek to understand the technology-driven extraction of racial capital as a multisited force that works across the above-mentioned scales while reconstituting borders continually.

GLOBAL MOVEMENT

The increasingly digital realities of global movement have changed practices of bordering, and therefore also the possibility for movement.

As movement is increasingly datafied and commodified, it becomes difficult to conceive of movement void of racial capitalist enclosure. Instead, the digital periphery has created the conditions under which refuge without fugitivity is the apparent reality. As will be clear from this section, migration scholars, particularly those of the abolitionist orientation, will have to contend with fugitivity from technology deployments as part and parcel of their analysis of refuge and movement going forward.

For example, Amazon, which has an extensive record of workers' rights abuses, has also been known to supply Amazon Web Services to provide cloud infrastructure for other technology products used by ICE.[13] ICE uses this software (e.g., Palantir's Gotham or Foundry products) to facilitate deportation, recommend detention, and assess immigrants' propensity to make positive contributions to society.[14] Microsoft, which also designed New York City's largest police video surveillance infrastructure (the Domain Awareness System), provides Azure Cloud Infrastructure services to ICE. Moreover, until 2020, the software giant held a 40 percent stake in controversial Israeli facial recognition startup, AnyVision (now known as Oosto).[15] Yet, the company also collaborates with Accenture, under the endorsement of the UNHCR, to develop digital ID systems for refugees. As actors who either directly supply both border control and/or humanitarian technologies in camps and cities of refuge, or indirectly fund initiatives designed in the name of refugees and newcomers, they convert migrant struggles into profit—an undeniable aspect of the digital periphery under racial capitalism. These are the very companies that have historically profited from utilizing the Global Majority countries as labs for untested experimental technology interventions. They continue to do so under a self-ascribed auspice of "tech for good."[16]

As made apparent through the genealogy of refugee and migrant reception in New York City and Berlin in chapter 3, there has been a convergence of the hostile immigration environment and Silicon

Valley logics[17]—logics that UN Special Rapporteur Tendayi Achiume, as recently as November 2020, warned were being deployed in the advancement of "xenophobic and racially discriminatory ideologies."[18] While the interactions between the technology sector and the racialized politics of migration appear unique and novel, they are undergirded by both a coloniality of migration[19] and the modes of subjugation central to the construction of racial capitalism.[20]

Humanitarians and technologists develop ICoTs based on their subjective and stereotypical perception of the modern refugee's plight. This creates computational perversions of refugee subjects, which are axiomatic representations of an enormously diverse populations of people. These representations are housed within the digital periphery, which flattens the essentialism of the population. This is an orientalism for the digital age, which can have material and potentially dire consequences for how migrant populations move and the spaces they can access. Fugitives will no longer be able to seek refuge—rather, moving bodies can move into the digital periphery, where their containment simply changes state.

ESCAPING THE DIGITAL PERIPHERY?

Reflecting on the involvement of tech companies in US immigration enforcement, my contact at Mijente succinctly describes Silicon Valley's market penetration: "It's their innovation curve: start with war, then refugees, then the mass." Campaigns such as Close the Camps, Mijente's #NoTechForICE, RethinkLinkNYC and the Immigrant Defense Project have taken direct action at the offices of technology giants such as Google, Amazon, and Palantir. They build on a rejection of the "smart" altogether. They present a break with approaches that seek to gradually improve the experience marginalized populations have with violent technology deployments, and that also assume the inevitability of tech. Their refusal—whether in the form of urban concealment, the use of alternative technology configurations, or physical tech sabotage—presents a critical neo-Luddite epistemology of migrant survival for life in digital cities of refuge.

In the year following the conclusion of my fieldwork, these efforts have begun to be referred to as abolitionist. In its 2020 report titled "Technologies for Liberation: Toward Abolitionist Futures," the Astraea Lesbian Foundation for Justice and partners, including Mijente, state:

> What distinguishes abolition as a strategy is that it does not assume that the use of carceral technologies and mass criminalization are inevitable. . . . If surveillance is, as one organizer put it, about "constant control of the body," then movements for abolition ask: How do we make structures of oppression and control irrelevant?[21]

In their nonuse and refusal, communities confer meaning and reveal partial truths about ICoTs and how such technologies perpetuate and expand the digital periphery. In so doing, they present the possibility for change, a vision in stark contrast and negation of techno-solutionism and techno-chauvinism.[22] Displaced communities, immigrants' rights advocates, and privacy activists, whom I encountered in both New York and Berlin (see chapters 4 and 5), have sought to challenge ICoTs deployed against migrants in particular, on grounds of their potentially devastating consequences for the communities such tools purportedly intend to serve.

I urge you to think about the current state of affairs as malleable; there's nothing fundamentally inevitable about surveillance, control, and entrapment. This is the core of abolitionism: acknowledging that these forces can and should be undone. Ours is also a moment that calls for a willingness to challenge the premise, and even existence, of certain technologies, and where necessary, to refuse them. To take on a necessarily neo-Luddite orientation has never been more important.

In conclusion, three key issues are worth our attention going forward. First, the increasing pervasiveness of the criminalization of antisurveillance activism is a problem, particularly as this works asymmetrically along lines of race. How is "techno-crime" formulated, framed, and disciplined in the digital periphery? President Lyndon Johnson's Cold War–era global "War on Crime" meant "treating the political dedication of revolutionaries as the permanence of crime and the incorrigibility of criminals."[23] Set against the backdrop of decolonization and the civil rights movement, this conflation between race, crime, and communism in law enforcement agencies meant the conflation of "crime with subversion." Today, neo-Luddite refusal of the kinds documented throughout chapters 4, 5, and 6 is a fundamental negation of the practices of control permitted through digital urban infrastructures. These practices of refusal, which range from subtler forms such as urban concealment to more direct actions like Mijente's actions against Palantir, Rodriguez's actions against LinkNYC kiosks, and New Yorkers' protests against OMNY readers during the J31 day, are already in the process of being criminalized. As such, under the conditions of the digital periphery, the criminalization of technological "deviance" begs the question: how can these systems be contested under the current sociolegal paradigm that is layered on racial capitalism?

Second, scholars must seek to understand and categorize the roster of technologies underpinning the interface between borders, cities, and the technology industry, including the networks that make up the software suites, the clouds, hardware, funders, and vendors. This has become especially urgent against the backdrop of the combination of a data-intensified COVID-19 pandemic,[24] exponentially increasing Islamophobic counterterrorism and security responses with their accompanying technical products, and more recent developments in artificial intelligence and Web3 at large. In November 2020, a *Vice* report uncovered that the US military had been purchasing "granular movement data of people around the world" from seemingly "innocuous" apps, such as a "Muslim prayer and Quran app . . ., a Muslim dating app, a popular Craigslist app, an app for following storms"—even a leveler app for DIY work.[25] These developments tell us that most, if not all, networked consumer technologies risk conscripting marginalized and racialized communities into the digital periphery to execute an illusion of control. This warrants an expansive investigation involving the above-mentioned roster of technologies, documenting how the increasing saturation of the digital periphery does away with what has come to be known as "function creep'" (when data intended for one purpose "exceeds its original purpose"),[26] situating the "creep" as an inextricable and deliberate "function" of ICoTs.

Finally, there is an urgent need to put in conversation the constituent parts of the digital periphery, such as across cities and other interstitial geographies such as camps, prisons, and detention centers. As explored in chapters 2 and 6, the digital periphery operates across prisms of borderization that are concerned less with geographic boundaries than marginal subjectivities and environments at large. Needless to say, the activities of technology actors increasingly follow the planetary rhythms of violence that animate global human movement as much between, within, across, and without borders. Neil Brenner and Christian Schmid's call for "new theoretical categories through which to investigate the relentless production and transformation of socio-spatial organisation across scales and territories" has, as such, never been more salient.[27] The task of comprehensively mapping the digital periphery "across scales and territories" promises to unveil how the military–, prison–, and border–industrial complexes connect and work together.[28]

This book has ultimately complicated celebratory narratives around technology deployments in marginalized contexts and communities on the move. It has sought to do so in solidarity and active participation with the communities concerned. The future research agenda suggested in these pages has been conceived of in continued solidarity and the hope that it will help unravel how race and alterity are constituted, innovated, and weaponized in service of racial capitalism, and how this can be resisted.

Notes

GLOSSARY

1. Hong, *Technologies of Speculation.*
2. Georgiou, "City of Refuge or Digital Order?," 602.
3. Omi and Winant, *Racial Formation in the United States.*
4. Robinson, *Black Marxism.*
5. Omi and Winant, *Racial Formation in the United States.*
6. See for instance Bonilla-Silva, *Racism without Racists.*
7. See also Green, *Smart Enough City.*
8. Browne, *Dark Matters.*
9. Mann, Nolan, and Wellman, "Sousveillance."

INTRODUCTION

1. Mijente, "Our DNA," accessed December 5, 2020, https://mijente.net/our-dna.

2. #NoTechForICE, accessed December 9, 2020, https://notechforice.com.

3. Mijente, Immigrant Defense Project, and National Immigration Project, "Who's behind ICE?"

4. Slisco, "What Is Palantir?"

5. McKenzie, "Refugees Don't Just Come to Nations"; Muggah and Abdenur, "Refugees and the City."

6. Maitland, *Digital Lifeline?*

7. Madianou, "Technocolonialism."

8. Chaar-López, "Sensing Intruders."

9. Biesecker, "Elbit America Develops AI-Based Solutions."

10. This is undergirded by the past decades' techno-liberalism—what Atanasoski and Vora refer to as "the political alibi of present-day racial capitalism" (*Surrogate Humanity*, 4).

11. Madianou, "Technocolonialism."

12. See UNHCR's reporting from the situation in Syria, as of 2023, available at https://reporting.unhcr.org/operational/situations/syria-situation#:~:text = Over%2012%20million%20Syrians%20remained,from%205.7%20million%20 in%202021.

13. See for instance Benton, "Smart Inclusive Cities."

14. See the former UN special rapporteur on contemporary forms of racism's report from September 2021, Achiume, "Racial and Xenophobic Discrimination."

15. Mirzoeff, "Artificial Vision, White Space and Racial Surveillance Capitalism."

16. Broussard, *Artificial Unintelligence*, 7.

17. Morozov, *To Save Everything, Click Here*, 8.

18. See for instance Arjun Appadurai's "colonial *imaginaire*" ("Number in the Colonial Imagination," 329).

19. Hindman, *Myth of Digital Democracy*.

20. IOM, "Missing Migrants Project," accessed June 28, 2024, https://missingmigrants.iom.int/data.

21. Deaths from 2003 to 2017. See ICE's own data, "List of Deaths in ICE Custody," last modified June 5, 2017, www.ice.gov/doclib/foia/reports/detaineedeaths-2003-2017.pdf; see also Shoichet, "Death Toll in ICE Custody."

22. Miller, "More Than a Wall."

23. Achiume, "Report of the Special Rapporteur."

24. Currier, "Prosecuting Parents."

25. Miroff, Goldstein, and Sacchetti, "'Deleted' Families."

26. Walia, *Undoing Border Imperialism.*

27. ICEwatch, accessed May 18, 2020, https://raidsmap.immdefense.org.

28. See, most prominently, the New Sanctuary Coalition, accessed December 11, 2020, www.newsanctuarynyc.org.

29. NYC.gov, "Mayor's Office of Immigrant Affairs Releases Third Annual Report."

30. Henri Lefebvre understands cities as crucial nodes for the realization of freedom, livelihood, and associational life: the right to the city "manifests itself as a superior form of rights: the right to freedom, to individualization in socialization, to habitat and to inhabit" (*Writings on Cities*, 174).

31. Frenkel, "Microsoft Employees Protest Work with ICE."

32. Couldry and Mejias, *Costs of Connection*, 5.

33. Noble, *Algorithms of Oppression.*

34. O'Neil, *Weapons of Math Destruction*; Eubanks, *Automating Inequality.*

35. Benjamin, *Race after Technology.*

36. See in particular Suarez-Villa, *Technocapitalism*; Cheney-Lippold, *We Are Data*; and Madianou, "Technocolonialism."

37. See for instance Green, *Smart Enough City.*

38. See in particular Georgiou, "*City of Refuge* or Digital Order?"

39. Zuboff, *Age of Surveillance Capitalism.*

40. See in particular Buolamwini and Gebru, "Gender Shades."

41. Cheney-Lippold, *We Are Data.*

42. Broussard, *More Than a Glitch.*

43. Zuboff, "Surveillance Capitalism and the Challenge of Collective Action."

44. Pasquale, *Black Box Society.*

45. Christin, "Ethnographer and the Algorithm."

46. Brayne, *Predict and Surveil.*

47. See for an instance Broeders and Hampshire, "Dreaming of Seamless Borders"; Amoore, "Algorithmic War"; Amoore, "Biometric Borders"; Balzacq, "Policy Tools of Securitization"; Broeders, "New Digital Borders of Europe"; Dijstelbloem, *Migration and the New Technological Borders in Europe*; Muller, "(Dis)Qualified Bodies"; Vaughan-Williams, "UK Border Security Continuum"; Dunleavy, *Digital Era Governance*; and Prins et al., *iGovernment.*

48. See also Cheesman, "Self-Sovereignty for Refugees?"; Molnar, "Technological Testing Grounds"; Latonero and Kift, "On Digital Passages and Borders"; and Villa-Nicholas, *Data Borders.*

49. Ali, "Decolonizing Information Narratives."

50. Mathers and Novelli, "Researching Resistance to Neoliberal Globalization."

51. Wasik, "Migrant Crisis Triggers a Wave of Tech Innovation."

CHAPTER 1

1. Villa-Nicholas, *Data Borders.*

2. United against Refugee Deaths, "List of 52,760 Documented Deaths of Refugees and Migrants Due to the Restrictive Policies of 'Fortress Europe,'" last modified June 7, 2023, http://unitedagainstrefugeedeaths.eu/wp-content/uploads/2014/06/ListofDeathsActual.pdf.

3. Qurashi, "Prevent Strategy and the UK 'War on Terror.'"

4. Ahmad, "9/11."

5. Qurashi, "Prevent Strategy."

6. Robinson, *Black Marxism.*

7. Frenkel, "Microsoft Employees Protest Work with ICE."

8. Mijente, "Palantir's Technology Used in Mississippi Raids."

9. Privacy International, "All Roads Lead to Palantir."

10. Pegg, "UK Awards Border Contract to Firm Criticised over Role in US Deportations."

11. Ahmed and Tondo, "Fortress Europe."

12. Robinson, *Black Marxism*, 11.

13. Robinson, 34.

14. Robinson, 34.

15. Stoler, *Imperial Debris.*

16. Heng, "Invention of Race in the European Middle Ages."

17. Quijano, "Coloniality of Power, Eurocentrism, and Latin America," 548.

18. Robinson, *Black Marxism*; Kelley, "What Did Cedric Robinson Mean by Racial Capitalism?'

19. Heng, "Invention of Race in the European Middle Ages I."

20. Fields and Fields, *Racecraft.*

21. Heng, "Invention of Race in the European Middle Ages," 259.

22. Heng, 259.

23. Robinson, *Black Marxism*.

24. Appadurai, "Number in the Colonial Imagination."

25. Appadurai, 317.

26. Appadurai, 317.

27. Appadurai, 314.

28. Omi and Winant, *Racial Formation in the United States*.

29. Black, *IBM and the Holocaust*.

30. In 1986, IBM ended thirty-four years of business operations in South Africa. See Walters, "IBM to End Its Presence in S. Africa." In 2015, the Electronic Frontier Foundation filed an amicus brief in a lawsuit filed by Black South Africans against IBM in the US Supreme Court for its history of enabling the national ID system that powered aspects of apartheid in South Africa. See Kayyali, "EFF Files Amicus Brief."

31. Quijano, "Coloniality of Power, Eurocentrism, and Latin America," 534.

32. Bennett, Foley, and Krebs, "Learning from the Past to Shape the Future."

33. Bennett, Foley, and Krebs.

34. Bennett, Foley, and Krebs.

35. Benjamin, *Race after Technology*, 100.

36. Robinson, *Black Marxism*, 26

37. Mirzoeff, "Artificial Vision, White Space and Racial Surveillance Capitalism."

38. Robinson, *Black Marxism*.

39. Walia, *Undoing Border Imperialism*; Tazzioli, "Containment through Mobility."

40. Walia, *Undoing Border Imperialism*.

41. Walia, 10.

42. Tazzioli, "Containment through Mobility," 2765.

43. Tazzioli.

44. Gilroy, *Black Atlantic*.

45. Di Maio, "The Mediterranean," 42.

46. Rajaram, "Refugees as Surplus Population."

47. Walia, *Undoing Border Imperialism*.

48. Rajaram, "Refugees as Surplus Population."

49. Smythe, "Black Mediterranean and the Politics of Imagination," 8.

50. Tazzioli, "Containment through Mobility."

51. Tazzioli, 2765.

52. Tazzioli and Garelli, "Containment beyond Detention"; also reflected in field notes taken during interviews with humanitarian workers in 2018.

53. Tazzioli and Garelli, 1020.

54. Frontex. "Artificial Intelligence-Based Capabilities."

55. Walia, *Undoing Border Imperialism.*

CHAPTER 2

1. Rostow, "Stages of Economic Growth."

2. ID2020, GHPC, and Digital Identity Alliance, accessed December 5, 2020, https://id2020.org.

3. "Humanitarian Aid," Iris Guard, accessed December 5, 2020, www.irisguard.com/industry-sectors/humanitarian-aid.

4. Worldcoin, accessed September 1, 2023, https://worldcoin.org.

5. Fejerskov, "New Technopolitics of Development and the Global South"; Thatcher, "Data Colonialism through Accumulation by Dispossession"; Zuboff, *Age of Surveillance Capitalism.*

6. Bhattacharya, "Hacking the Refugee Crisis."

7. Morozov, *To Save Everything, Click Here.*

8. Wasik, "Migrant Crisis Triggers a Wave of Tech Innovation."

9. "Google x Techfugees Copenhagen Hackathon—March," Eventbrite, accessed June 30, 2024, www.eventbrite.com/e/google-x-techfugees-copenhagen-hackathon-march-tickets-90626978733.

10. Wasik, "Migrant Crisis Triggers a Wave of Tech Innovation."

11. Chan, *Networking Peripheries.*

12. Muselli et al., "Open-Source World Is More and More Closed."

13. Chan, *Networking Peripheries*, 118.

14. Chan, 118.

15. Chan, 118.

16. Hoffmann, Proferes, and Zimmer, "Making the World More Open and Connected."

17. Toyama, *Geek Heresy*, 65.

18. Green, *Smart Enough City*.

19. Geraghty et al., "Future of Cities."

20. As we've seen through the use of the SenseTime system in Hangzhou, Sanmenxia, and Wenzhou in China; see Mozur, "One Month, 500,000 Face Scans."

21. See Amnesty International's "Automated Apartheid" report on how facial recognition is used to automate violent restrictions on the freedom of movement in Palestine, reinforcing apartheid.

22. Zukin, *Innovation Complex*.

23. NYC.gov, "De Blasio Administration Announces NYCx."

24. Shapiro, *Design, Control, Predict*, 11.

25. Bruce Sterling quoted in Shapiro, 152.

26. Kelly, "Facial Recognition Smartwatches to Be Used."

27. Orwell, *1984*.

28. Williams, *Stand Out of Our Light*.

29. Alegre, "Rethinking Freedom of Thought for the 21st Century."

30. Wallerstein, "Modern World-System as a Capitalist World-Economy"; Amin, "Accumulation and Development"; Rostow, "Stages of Economic Growth."

31. As Juan Llamas Rodriquez argues, borders themselves are constructed visually through multimedia representations, layering them with symbolic meaning beyond their physical demarcation. It stands to reason that border subjects, then, would face the same form of abstraction. See more in Llamas-Rodriguez, *Border Tunnels*.

32. Benjamin, *Race after Technology*.

33. Benjamin, 119.

34. Worldcoin, "Worldcoin Begins Rollout of 1.5k Orbs to Meet Global Demand."

35. Worldcoin, "Non-profit and Refugee Aid Distribution," accessed November 15, 2023, https://docs.worldcoin.org/use-cases/wealth-distribution.

36. Noestlinger, Baehr, and Howcroft, "Worldcoin Says Will Allow Companies."

37. Betterplace Lab, "Refugee Tech."

38. In particular, the EURODAC and EUROSUR systems are used as extension of Frontex; see Latonero and Kift, "On Digital Passages and Borders."

39. Tazzioli, "How Cashless Programmes to Support Refugees' Independence."

40. Dijstelbloem and Meijer, *Migration and the New Technological Borders in Europe.*

41. Broeders and Hampshire, "Dreaming of Seamless Borders."

42. Friedman, "Mathematical Equations That Could Decide the Fate of Refugees"; Altemeyer-Bartscher et al., "On the Distribution of Refugees in the EU."

43. Broeders and Hampshire, "Dreaming of Seamless Borders."

44. UNHCR Innovation, "Chasing Opportunity."

45. UNHCR Innovation.

46. Schmitt et al., "Cellular and Internet Connectivity for Displaced Populations."

47. Stenum, "Body-Border."

48. Stenum.

49. Kingston, "Biometric Identification, Displacement, and Protection Gaps."

CHAPTER 3

1. Price and Benton-Short, "Counting Immigrants in Cities across the Globe."

2. Price and Benton-Short.

3. Foner, "Immigration History and the Remaking of New York," 29. See also Foner, *New Immigrants in New York*; and Foner et al., *New York and Amsterdam.*

4. Glaeser, "Urban Colossus," 18.

5. Foner, "Immigration History and the Remaking of New York," 42.

6. Amnesty International, "Licence to Discriminate."

7. In New York, the USCRI serves as the VolAg in charge of R&P.

8. Kasinitz, *Caribbean New York.*

9. Field note, Anita, December 2018, Brooklyn.

10. Field note, Anita.

11. Field note, Anita.

12. Field note, Dino, December 2018, Brooklyn.

13. Field note, Dino.

14. Interview, Yunus, November 2018, Brooklyn.

15. Someone arriving as a refugee without a connection in the United States.

16. Someone arriving as a refugee with a connection in the United States.

17. Field note, Abdo, December 2018, Brooklyn.

18. Field note, Anita, December 2018, Brooklyn.

19. I worked with Reset during the four months between New York being determined as an area of concern and the organization's evaluation for continuing R&P services.

20. Reflective note, Reset, November 2018, Brooklyn.

21. Lazarus, "New Colossus," 752.

22. Thomsen, "Comey Quotes Statue of Liberty Poem."

23. Bayor, *Encountering Ellis Island.*

24. Markel and Stern, "Foreignness of Germs."

25. Johnson and Lubin, *Futures of Black Radicalism.*

26. Markel and Stern, "Foreignness of Germs."

27. At the time of writing, the COVID-19 pandemic continued to be used as justification for racialized xenophobic discourse. See also Aimone, "1918 Influenza Epidemic in New York City."

28. Markel and Stern, "Foreignness of Germs."

29. Zucker, "Refugee Resettlement in the United States."

30. Brown and Scribner, "Unfulfilled Promises, Future Possibilities."

31. Markel and Stern, "Foreignness of Germs."

32. United States Holocaust Memorial Museum, "United States Immigration and Refugee Law, 1921–1980," accessed November 24, 2020, https://encyclopedia.ushmm.org/content/en/article/united-states-immigration-and-refugee-law-1921-1980.

33. Bateman-House and Fairchild, "Medical Examination of Immigrants at Ellis Island."

34. Refugee Act of 1980, Pub. L. No. 96–212, 94 Stat. 102 (1980), www.govinfo.gov/content/pkg/STATUTE-94/pdf/STATUTE-94-Pg102.pdf.

35. Darrow, "(Re)Construction of the U.S. Department of State's Reception and Placement Program."

36. Brady, "German Right-Wing Populist Movement Put under Surveillance"; Reuters, "Germany Steps up Warnings about Right-Wing Identitarian Movement."

37. Heng, "Invention of Race in the European Middle Ages."

38. Altemeyer-Bartscher et al., "On the Distribution of Refugees in the EU."

39. Rhoades, "Foreign Labor and German Industrial Capitalism," 553.

40. Gesley, "Germany."

41. Anil, "No More Foreigners?," 460

42. Anil.

43. Brady, "German Right-Wing Populist Movement Put under Surveillance."

44. European Commission, "Digital Skills Gap in Europe."

45. Sánchez Nicolás, "Digital Gap."

46. Buck, "German Start-Ups Attract Record Investment in 2017."

47. Buck.

48. Turk, "UK's Tech Sector Faces a Tougher Talent Battle Post-Brexit."

49. Lewicki, "Race, Islamophobia and the Politics of Citizenship in Post-Unification Germany."

50. Rist, "Migration and Marginality."

51. Gesley, "Germany."

52. Anil, "No More Foreigners?"

53. Chin, *Guest Worker Question in Postwar Germany*, 32.

54. Rist, "Migration and Marginality," 96.

55. "Ethnic German resettlers are considered Germans within the meaning of article 116, paragraph 1 of the German Basic Law. They are defined as people of German heritage from the successor states of the former Soviet Union and from other Eastern European States, as well as China. Further requirements for acquiring 'ethnic German resettler' status are that they were born before January 1, 1993; resided in the described territories either since the end of World War II, since March 31, 1952, or since their birth if a parent met one of the two record dates; left the described territories after December 31, 1992; submitted an application for recognition as an 'ethnic German resettler'; and took up permanent residence in Germany within six months of leaving the designated territories. People who are not from one of the successor states of the Soviet Union must also prove that they suffered disadvantages or discrimination because of their German heritage. Once someone has been recognized as an 'ethnic German resettler,' he or she is automatically awarded German citizenship" (Gesley, "Germany").

56. Gesley.

57. Gesley.

58. Rhoades, "Foreign Labor and German Industrial Capitalism," 553.

59. Gesley, "Germany."

60. By 1977, nearly 12 percent of the Federal Republic's labor force consisted of guest workers, with Berlin ranking as the "city with the fifth largest Turkish population in the world" (Rist 1979)

61. Rist, "Migration and Marginality."

62. Rist, 98.

63. "Requirements were that the guest workers could not be married to a German citizen; lost his or her job because the business or the main components of the business had been shut down or had gone bankrupt; had applied for return assistance by June 30, 1984; had been legally residing in Germany until the date of departure; and had permanently left Germany with his or her family between October 30, 1983, and September 30, 1984" (Gesley, "Germany").

64. Gesley, "Germany."

65. Rist, "Migration and Marginality," 99.

66. Gesley, "Germany."

67. Gesley.

68. Gesley, in particular: "The 'airport procedure' applied to asylum seekers from 'safe countries of origin' and to those who did not have a passport or other valid travel documents upon arrival at the airport. Under the procedure, the asylum seeker stayed in the transit area and a decision whether to grant him or her entry to the territory and to the general asylum procedure was made under an expedited procedure. If the immigration officer found that the application was 'manifestly unfounded,' the applicant was denied entry to the territory and deportation was threatened as a precautionary measure."

69. Anil, "No More Foreigners?," 458, noting that in 1991, the Maastricht Treaty granted local election voting rights across EU member states.

70. Lewicki, "Race, Islamophobia and the Politics of Citizenship."

71. Lewicki, 508.

72. Gesley, "Germany."

73. Children born to foreign families would now be eligible for German citizenship and provided legal residence in the country for at least eight years (Gesley).

74. The Residence Act simplified residency titles to "temporary residence" and "permanent settlement" permits (Gesley).

75. Gelsey.

76. Lewicki, "Race, Islamophobia and the Politics of Citizenship."

77. Reportedly over a thousand men, characterized as "North African" and "Arabs," engaged in theft and groping over the course of New Year's Eve, catalyzing parliamentary proceedings that would tighten and further complicate asylum processes going forwards. See Weber, "German Refugee 'Crisis' after Cologne."

78. Lewicki, "Race, Islamophobia and the Politics of Citizenship," 508.

79. Robinson, *Black Marxism*.

80. Lewicki, "Race, Islamophobia and the Politics of Citizenship."

81. De Genova, "Spectacles of Migrant 'Illegality.'"

82. Gutierrez Rodriguez, "Coloniality of Migration and the 'Refugee Crisis'"; Quijano, "Coloniality of Power, Eurocentrism, and Latin America."

83. Lewicki, "Race, Islamophobia and the Politics of Citizenship."

84. Lewicki.

85. Gesley, "Germany."

86. Mischke, "Germany Passes Controversial Migration Law."

87. Mischke.

88. Rist, "Migration and Marginality."

89. Rhoades, "Foreign Labor and German Industrial Capitalism."

90. Harvey, "'New' Imperialism."

CHAPTER 4

1. Field note, Maryam, November 2018, Brooklyn.

2. Rush, "'Private' Refugee Resettlement Agencies."

3. Interview, Amelia and Melissa, November 2018, Manhattan.

4. Currier, "Prosecuting Parents."

5. Miroff, Goldstein, and Sacchetti, "'Deleted' Families."

6. Walia, *Undoing Border Imperialism*.

7. Davis and Akers Chacón, *No One Is Illegal*, 15.

8. McFarlane and Söderström, "On Alternative Smart Cities."

9. Alves, *Anti-Black City*.

10. The area gained particular notoriety for interdiaspora conflict some three decades earlier in the summer of 1991, leading to severe tensions between the Hasidic Jewish community and the Afro-American and Afro-Caribbean communities. See Buff, "Teaching Crown Heights."

11. New York University, "Crown Heights from the 1950s to Today," accessed May 19, 2020, https://wp.nyu.edu/crownheights/history-and-geography/crown-heights-from-the-1950s-to-today.

12. Foner, "Immigration History and the Remaking of New York."

13. Devereaux, "How ICE Operations in New York Set the Stage."

14. Interview, David, September 2018, Brooklyn; see also Eisenzweig, "In the Shadow of Child Protective Services."

15. Eisenzweig.

16. Dreby, "How Today's Immigration Enforcement Policies Impact Children, Families, and Communities."

17. Muñiz, *Borderland Circuitry.*

18. McKenzie, "Refugees Don't Just Come to Nations"; Muggah, "Safe Havens."

19. Field note, support group meeting, Majeed, October 2018, Manhattan.

20. Field note, support group meeting, Abbas and Talal, October 2018, Manhattan.

21. Interview, David, September 2018, Brooklyn.

22. Interview, David, September 2018, Brooklyn.

23. Pierce and Bolter, "Dismantling and Reconstructing the U.S. Immigration System."

24. For instance, Sarah Lamdan has exposed how "data cartels," consisting of seemingly innocuous "data analytics" providers, link everything from library information and outdated online data with law enforcement (*Data Cartels*).

25. Interview, David, September 2018, Brooklyn.

26. Field note, Anita, December 2018, Brooklyn.

27. Sidewalk Labs is a sister company of Google, both of which are owned by Alphabet.

28. Wiggers, "LinkNYC's 5 Million Users Make 500,000 Phone Calls Each Month."

29. Kofman, "Are New York's Free LinkNYC Internet Kiosks Tracking Your Movements?"

30. Kofman.

31. Interview, Isabel, March 2019, online.

32. Interview, Mike, March 2019, online.

33. Interview, Zeynep, March 2019, online.

34. Interview, Isabel, March 2019, online.

35. Field note, Alhasan, January 2019, Brooklyn.

36. Interview, Yunus, November 2019, Brooklyn.

37. People's Forum, accessed June 30, 2024, https://peoplesforum.org.

38. Interview, Amelia and Melissa, November 16, 2018, Manhattan.

39. Interview, Amelia and Melissa, November 16, 2018, Manhattan.

40. Interview, Amelia and Melissa, November 16, 2018, Manhattan.

41. Broussard, *Artificial Unintelligence.*

42. Interview, Amelia and Melissa, November 16, 2018, Manhattan.

43. Alluding to Doctoroff's urban planning memoir *Greater Than Ever,* Amelia points to his professed hatred for the city growing up. In an interview from 2017, Doctoroff admits that his view of the city became more "romanticized" as he started seeing New York through the eyes of "our competitors" while working on its Olympic bid (Plitt, "Q&A").

44. NYC.gov, "De Blasio Administration Announces Winner of Competition."

45. Pinto, "Google Is Transforming NYC's Payphones."

46. CNN, "DeBlasio Defends NY Sanctuary City Policies."

47. NYC.gov, "De Blasio Administration Announces NYCx."

48. Zukin, *Innovation Complex.*

49. NYC.gov, "On Citizenship Day, de Blasio Administration Launches New Campaign."

50. Brook, *Unruly Cities?*, 130.

51. Tulumello, "From 'Spaces of Fear' to 'Fearscapes.'"

52. Gandy, *Panoptic Sort*, 15.

53. Interview, Isabel, March 2019, online.

54. Interview, Yunus, November 2019, Brooklyn.

55. *Manhattan Times News.* "City Unveils New IDNYC Features."

56. IDNYC, accessed March 9, 2020, https://www1.nyc.gov/site/idnyc/index.page.

57. Field note, Georgiana, November 2018, Brooklyn.

58. Interview, Grisha, November 2018, Manhattan.

59. Interview, Yunus, November 2019, Brooklyn.

60. Reviews posted to IDNYC Facebook page, accessed October 7, 2018, www.facebook.com/IDNYCForAll.

61. Robbins, "He Delivered Pizza to an Army Base in Brooklyn."

62. Greenberg, "IDs Were Meant to Protect Immigrants."

63. Field note, Jane, September 2019, online.

64. NYCLU, "Testimony of the New York Civil Liberties Union."

65. Sanders, "Council Bill May Block Smart Chips in New York Municipal IDs."

66. Sanders.

67. New Economy Project, "Letter to Mayor Bill de Blasio on Proposal."

68. Miller, "Cubic Wins $500M Contract to Overhaul NYC Subway and Bus Fare System."

69. Since 2003 Cubic has also operated the London Oyster card system, which works through an RFID-enabled "tap and go" card. It has since been upgraded to support contactless credit cards and other smart devices with NFC support (initially through the Apple Pay and Google Pay services).

70. Surveillance Technologies Oversight Project, "OMNY Surveillance Oh My."

71. "OMNY Privacy Policy," last modified June 7, 2022, https://omny.info/privacy-policy.

72. Chung, "MTA's New OMNY Scanners Have Cameras in Them."

73. OMNY, "If You See Something, Say Something," accessed June 30, 2024, https://omny.info/see-something-say-something.

74. *Spectrum News*, "NYPD Seeks Woman Seen Smashing OMNY Screens with Hammer."

75. Field note, Yunus, December 2018, Brooklyn.

76. The initiative intended to remove significant bureaucratic hurdles. Jennifer, an informant from the HPD, explained to me on August 23, 2019, that in the past the process was completely manual. Applicants would submit an envelope containing the application and receive a log number at the noted address. Applications were subsequently all put into a trash bag and shuffled around, before anyone at the housing department was allowed to look at it. This led to several deficiencies: applications were lost, there was no guarantee that had a

legitimate shuffle had occurred, and the paperwork alone was hard to manage, causing major delays. See Plitt, "How to Apply for NYC's Affordable Housing Lottery."

77. Plitt.

78. MOIA, "Housing," accessed March 9, 2020, www.nyc.gov/site/immigrants/city-services/housing.page.

79. Interview, Abigail, March 2020, online.

80. Field note, Yunus, December 2018, Brooklyn.

81. Interview, David, September 2018, Brooklyn.

82. Field note, Alex, December 2018, Brooklyn.

83. Field note, support group meeting, Farukh, October 2018, Manhattan.

84. Field note, support group meeting, Farukh, October 2018, Manhattan.

85. Field note, support group meeting, Abbas and Talal, October 2018, Manhattan.

86. Interview, Grisha, November 2018, Manhattan.

87. Interview, Anneliese and Erin, September 2018, Manhattan.

88. Interview, Anneliese, August 2019, online.

89. Interview, Mark, September 2018, Brooklyn.

90. Interview, Mark, September 2018, Brooklyn.

91. Weissert and Miller, "Mexico Agrees to Invest $1.5B."

CHAPTER 5

1. Interview, Nadim, May 2019, Berlin.

2. Betterplace Lab, "Refugee Tech."

3. Betterplace Lab.

4. Betterplace Lab.

5. Brock, "Critical Technocultural Discourse Analysis."

6. Smith, "Shocking Images of Drowned Syrian Boy Show Tragic Plight of Refugees."

7. Guibernau, "Migration and the Rise of the Radical Right."

8. Danewid, "White Innocence in the Black Mediterranean," 1684.

9. Erel et al., "Understanding the Contemporary Race–Migration Nexus."

10. Interview, Nadim, May 2019, Berlin.

11. Interview, Matteo, April 2019, Berlin.

12. As discussed in chapter 3, Alternative für Deutschland is a far-right anti-immigrant political party in Germany.

13. Field note, Anna, April 2019, Berlin.

14. Interview, Nadim, May 2019, Berlin.

15. Wallis, "Not Good for the Economy."

16. Interview, Sarina, May 2019, Berlin.

17. Field note, G100 prep day, April 2019, Berlin.

18. Deleuze, "Postscript on the Societies of Control"; Cheney-Lippod, *We Are Data*.

19. Interview, Sarina, May 2019, Berlin.

20. Brock, "Critical Technocultural Discourse Analysis."

21. "Germany from A to Z," Handbook Germany, accessed July 1, 2024, https://handbookgermany.de/en.

22. Field note, Hakan, May 2019, Berlin.

23. O'Malley, "Stop Acting Surprised That Refugees Have Smartphones."

24. AbuJarour et al., "Your Home Screen Is Worth a Thousand Words."

25. Interview, Edin, April 2019, Berlin.

26. Georgiou, "Does the Subaltern Speak?," 49.

27. Interview, Nadim, May 2019, Berlin.

28. Terranova, *Network Culture*.

29. Mohler, "Just Do It."

30. Vincent, "Forty Percent of 'AI Startups' in Europe Don't Actually Use AI."

31. Note for instance also how surveillance companies like Palantir and Clearview claim that their products can be used to find trafficked migrants, without ever disclosing any evidence supporting such grand postures. "How Facial Recognition Is Identifying Human Trafficking Victims," Clearview AI, accessed June 28, 2024, www.clearview.ai/post/how-facial-recognition-is-identifying-human-trafficking-victims; "Fighting Child Exploitation with Big Data," Palantir, accessed June 28, 2024, www.palantir.com/ncmec. Meanwhile, there is plenty of evidence attesting to how these products have been used to subject communities on the move to surveillance, violence, and removal. See also Georgiou, "*City of Refuge* or Digital Order?"; and de Laat and Gorin, "Iconographies of Humanitarian Aid in Africa."

32. Goodfellow, *Hostile Environment.*

33. Ilse, "Royal Germany Tour Recap"; International Rescue Committee, "Prince of Wales and the Duchess of Cornwall Meet Refugee Women in Berlin."

34. Interview, Tom, May 2019, Berlin.

35. Interview, Tom, May 2019, Berlin.

36. Interview, Tom, May 2019, Berlin.

37. Interview, Kelsey, January 2020, online.

38. Interview, Maddie and Rebecca, February 2020, online.

39. Interview, Yanos, July 2019, online.

40. One Young World, "How This Entrepreneur Is Using Blockchain for Humanitarian Projects."

41. Mastercard, "Smart Communities Coalition," accessed June 28, 2024, www.mastercard.us/content/dam/public/mastercardcom/na/us/en/governments/others/scc-overview-fact-sheet-Oct-10-2019.pdf.

42. Benjamin, *Race after Technology.*

43. "Wer wir sind," Gesellschaft für Freiheitsrechte, accessed July 1, 2024, https://freiheitsrechte.org/ueber-die-gff/werwirsind.

44. Turß, "Invading Refugees' Phones."

45. Interview, Nadim, May 2019, Berlin.

46. Field note, Hakan, May 2019, Berlin.

47. Interview, Edin, April 2019, Berlin.

48. Interview, Nadim, May 2019, Berlin.

49. Interview, Sarina, May 2019, Berlin.

50. Interview, Yara, May 2019, Berlin.

51. Interview, Yara, May 2019, Berlin.

52. Pruchniewska, "Group That's Just Women for Women."

53. Brock, "Critical Technocultural Discourse Analysis."

54. Appadurai, *Modernity at Large*; De Genova, "Spectacles of Migrant 'Illegality.'"

55. Heng, *Invention of Race in the European Middle Ages*, 20.

56. Heng.

57. Brock, "Critical Technocultural Discourse Analysis."

58. Benjamin, *Race after Technology.*

CHAPTER 6

1. Mbembe, "Bodies as Borders," 9.

2. Maynard, "Black Life and Death across the U.S.–Canada Border."

3. Mezzadra, "Abolitionist Vistas of the Human."

4. Maynard, "Black Life and Death across the U.S.–Canada Border," 215.

5. Pasquale, *Black Box Society*; Christin, "Ethnographer and the Algorithm."

6. Christin, 906.

7. Easterling, *Extrastatecraft*.

8. Christin, "Ethnographer and the Algorithm," 913.

9. "The process by which certain spaces are transformed into uncrossable places for certain classes of populations, who thereby undergo a process of racialization" (Mbembe, "Bodies as Borders," 9).

10. Interview, Anita, December 2018, New York.

11. Quijano, "Coloniality of Power, Eurocentrism, and Latin America."

12. Abujarour et al., "Your Home Screen Is Worth a Thousand Words."

13. Tsibolane and Brown, "Principles for Conducting Critical Research Using Postcolonial Theory."

14. Madianou, "Technocolonialism."

15. Quijano, "Coloniality of Power, Eurocentrism, and Latin America."

16. Practices ranging from smartphone vendor reliance on the "exploitation of Chinese workers at Foxconn factories" and e-waste dumping in poorer countries including Ghana, right through to the provision of a rudimentary limited (but free) version of the internet by Facebook's Free Basics program (Madianou, "Technocolonialism") and the deployment of digital urban infrastructures in New York City, are all connected through the digital periphery.

17. Fejerskov, "New Technopolitics of Development and the Global South."

18. Granqvist, "Assessing ICT in Development," 292.

19. Field note, Anna, April 2019, Berlin.

20. de Laat and Gorin, "Iconographies of Humanitarian Aid in Africa."

21. Omi and Winant, *Racial Formation in the United States*; Robinson, *Black Marxism*.

22. Franklin, *Control*, 4.

23. Georgiou, "*City of Refuge* or Digital Order?," 602.

24. Terranova, *Network Culture*, 75.

25. Lefebvre, *Production of Space*.

26. Field note, Isabel, March 2019, online.

27. As per Jonathon Penney, the "chilling effect doctrine" was first defined as an encouragement to "courts to treat rules or government actions that 'might deter' the free exercise of First Amendment rights 'with suspicion'" ("Chilling Effects," 125).

28. Benjamin, *Race after Technology*, 89.

29. Interview, Amelia and Melissa, November 16, 2018.

30. Somewhat tangentially and yet bizarrely, at the time and place of writing, the same hardware provided by Intersection for LinkNYC has emerged across London, warning the good people of Hackney that the risk of COVID-19 transmission in their particular area is "high."

31. Garner, "European Union and the Racialization of Immigration."

32. Interview, Tom, May 2019, Berlin.

33. Martin, "Slavery's Invisible Engine," 819.

34. Mirzoeff, "Artificial Vision, White Space and Racial Surveillance Capitalism," 1296.

35. Easterling, *Extrastatecraft*.

36. Easterling.

37. NYC.gov, "De Blasio Administration Announces NYCx."

38. Zukin, *Innovation Complex*.

39. Easterling, *Extrastatecraft*.

40. Costanza-Chock, *Design Justice*, 82.

41. Interview, Stan, March 2019, Berlin.

42. Costanza-Chock, "Design Justice."

43. Easterling, *Extrastatecraft*.

44. Marcuse, "Enclave, the Citadel, and the Ghetto."

45. Shapiro, *Design, Control, Predict*, 11.

46. Shapiro.

47. Bruce Sterling quoted in Shapiro, 152.

48. S. C. quoted in Hickman in Wei and Peters, "'Intelligent Capitalism' and the Disappearance of Labour," 757.

49. Zuboff, "Surveillance Capitalism and the Challenge of Collective Action," 11.

50. Zuboff, 13.

51. Brock, "Critical Technocultural Discourse Analysis."

52. See also Hong, *Technologies of Speculation*, for a more granular analysis of datafication, adding further dimension to my elaboration of surrogate data.

53. Morozov, "Digital Intermediation of Everything."

54. Cheney-Lippold, *We Are Data*.

55. Benjamin, *Race after Technology*.

56. This builds on Miranda Fricker's notion of testimonial injustice, a perceived deficit in the credibility of an individual due to physical attributes and characteristics related to their identity (*Epistemic Injustice*).

57. This, on the other hand, builds on Fricker's notion of hermeneutical injustice, the social condition under which the knowledge needed to act in the best self-interest of an individual is withheld (*Epistemic Injustice*).

58. OpenAI, "Introducing ChatGPT."

59. Interview, Sarina, May 2019, Berlin.

60. Srinivasan and Pyati, "Diasporic Information Environments"; Ponzanesi, "Digital Diasporas"; Georgiou, *Diaspora, Identity, and the Media*; Appadurai, *Modernity at Large*.

61. Benjamin, *Race after Technology*, 34.

62. Cedric Robinson discussed in Johnson and Lubin, *Futures of Black Radicalism*, 36.

63. Benjamin, *Race after Technology*, 34.

64. Trueman, "Light That Never Goes Out."

65. Glendinning, "Notes toward a Neo-Luddite Manifesto."

66. Glendinning.

67. Glendinning.

68. Hill, "Violent Opt-Out."

69. First presented at the Intercultural Digital Ethics Symposium at the University of Oxford in October 2019.

70. Fanon, *Wretched of the Earth*, 42.

CONCLUSION

1. Benton, "Smart Inclusive Cities."

2. Morozov, *To Save Everything, Click Here*; Broussard, *Artificial Unintelligence*.

3. Mahmoudi, "Digital Borders in [Smart] Cities of Refuge."

4. Mahmoudi.

5. Stenum, "Body-Border."

6. Weitzberg, "Machine-Readable Refugees."

7. Weitzberg.

8. European Commission, "Horizon 2020," accessed September 1, 2023, https://ec.europa.eu/programmes/horizon2020/en.

9. Campbell, "Sci-Fi Surveillance"; Ahmed, "EU Accused of Abandoning Migrants to the Sea"; Molnar, "Technological Testing Grounds."

10. Mozur, "One Month, 500,000 Face Scans"; Brandom, "Huawei Worked on Facial Recognition System to Surveil Uighurs."

11. Duan et al., "Expression of Concern," 125.

12. Field note, Maddie and Rebecca, April 2019, online.

13. Mijente, Immigrant Defense Project, and National Immigration Project, "Who's behind ICE?"

14. McCarroll, "Weapons of Mass Deportation."

15. Dastin, "Microsoft to Divest AnyVision Stake."

16. Fejerskov, "New Technopolitics of Development and the Global South"; Madianou, "Technocolonialism."

17. Mbembe, "Bodies as Borders"; Georgiou, "*City of Refuge* or Digital Order?"; Robinson, *Black Marxism*.

18. Achiume, "Report of the Special Rapporteur."

19. Gutierrez Rodriguez, "Coloniality of Migration and the 'Refugee Crisis.'"

20. Robinson, *Black Marxism*.

21. Emmer et al., "Technologies for Liberation."

22. Morozov, *To Save Everything, Click Here*; Broussard, *Artificial Unintelligence*.

23. Schrader, *Badges without Borders*, 50.

24. Taylor et al., *Data Justice and COVID-19*.

25. Cox, "How the U.S. Military Buys Location Data from Ordinary Apps."

26. Jacobsen, "Experimentation in Humanitarian Locations."

27. Brenner and Schmid, "Planetary Urbanisation," 13.

28. Eisenhower, "Military–Industrial Complex Speech"; Davis, *Prison Industrial Complex*; Miller, "More Than a Wall."

Bibliography

AbuJarour, Safa'a, Cora Bergert, Jana Gundlach, Antonia Köster, and Hanna Krasnova. "'Your Home Screen Is Worth a Thousand Words': Investigating the Prevalence of Smartphone Apps among Refugees in Germany." *AMCIS 2019 Proceedings* 3 (2019): https://aisel.aisnet.org/amcis2019/social_inclusion/social_inclusion/3.

Access Now. "The Toronto Declaration: Protecting the Rights to Equality and Non-Discrimination in Machine Learning Systems." May 16, 2018. www.accessnow.org/the-toronto-declaration-protecting-the-rights-to-equality-and-non-discrimination-in-machine-learning-systems.

Acharya, Sanghmitra S., Sucharita Sen, Milap Punia, and Sunita Reddy. *Marginalization in Globalizing Delhi: Issues of Land, Livelihoods and Health*. New Delhi: Springer India, 2016.

Achiume, Tendayi. "Racial and Xenophobic Discrimination and the Use of Digital Technologies in Border and Immigration Enforcement—Report of the Special Rapporteur on Contemporary Forms of Racism, Racial Discrimination, Xenophobia and Related Intolerance." UN Human Rights Office, September 22, 2021.

www.ohchr.org/en/documents/thematic-reports/ahrc4876-racial-and-xenophobic-discrimination-and-use-digital.

———. "Report of the Special Rapporteur on Contemporary Forms of Racism, Racial Discrimination, Xenophobia and Related Intolerance." UN Human Rights Office, 2020. https://antiracismsr.org/wp-content/uploads/2020/11/A_75_590_Advance-Unedited-Version.pdf.

Adibifar, Karam. "Technology and Alienation in Modern-Day Societies." *International Journal of Social Science Studies* 4, no. 9 (2016): https://doi.org/10.11114/ijsss.v4i9.1797.

Ahad Divia Mattoo, Corinne Goldberg, Jillian Johnson, Carolina Farias Riaño, and Aliyyah Ahad. "Immigrants in the Smart City: The Potential of City Digital Strategies to Facilitate Immigrant Integration." Migration Policy Institute, November 18, 2015. www.migrationpolicy.org/article/immigrants-smart-city-potential-city-digital-strategies-facilitate-immigrant-integration.

Ahmad, Jared. "9/11: How Politicians and the Media Turned Terrorism into an Islamic Issue." *The Conversation*, September 10, 2021. https://theconversation.com/9-11-how-politicians-and-the-media-turned-terrorism-into-an-islamic-issue-167733.

Ahmed, Kaamil. "EU Accused of Abandoning Migrants to the Sea with Shift to Drone Surveillance." *The Guardian*, October 28, 2020. www.theguardian.com/global-development/2020/oct/28/eu-accused-of-abandoning-migrants-to-the-sea-with-shift-to-drone-surveillance.

Ahmed, Kaamil, and Lorenzo Tondo. "Fortress Europe: The Millions Spent on Military-Grade Tech to Deter Refugees." *The Guardian*, December 6, 2021. www.theguardian.com/global-development/2021/dec/06/fortress-europe-the-millions-spent-on-military-grade-tech-to-deter-refugees.

Aimone, Francesco. "The 1918 Influenza Epidemic in New York City: A Review of the Public Health Response." *Public Health Reports* 125, no. S3 (2010): 71–79. https://doi.org/10.1177/00333549101250S310.

Ajanaku, Lucas. "NCC Tackles Google over Free Wi-Fi." *Nation Newspaper*, February 21, 2019. https://thenationonlineng.net/ncc-tackles-google-over-free-wi-fi.

Alcantara, Christopher, and Caroline Dick. "Decolonization in a Digital Age: Cryptocurrencies and Indigenous Self-Determination in Canada." *Canadian Journal of Law and Society / Revue Canadienne Droit et Société* 32, no. 1 (April 2017): 19–35. https://doi.org/10.1017/cls.2017.1.

Alegre, Susie. "Rethinking Freedom of Thought for the 21st Century." *European Human Rights Law Review*, no. 3 (2017): 221–33. http://doi.org/10.13140/RG.2.2.27905.07529.

Ali, Syed Mustafa. "A Brief Introduction to Decolonial Computing." *XRDS: Crossroads, the ACM Magazine for Students* 22, no. 4 (2016): 16–21. https://doi.org/10.1145/2930886.

———. "Decolonizing Information Narratives: Entangled Apocalyptics, Algorithmic Racism and the Myths of History." *Proceedings* 1, no. 3 (2017): 50. https://doi.org/10.3390/IS4SI-2017-03910.

———. "'White Crisis' and/as 'Existential Risk,' or the Entangled Apocalypticism of Artificial Intelligence." *Zygon* 54, no. 1 (2019): 207–24. https://doi.org/10.1111/zygo.12498.

Allen, Samantha Wright. "Feds Tout 'Steady Progress' on Closing Data, Service Gaps Exposed during Syrian Refugee Operation." *Hill Times*, September 12, 2018. www.hilltimes.com/2018/09/12/steady-progress-data-settlement-service-gaps-emerged-operation-syrian-refugee/168512.

Alston, Philip, and Colin Gillespie. "Global Human Rights Monitoring, New Technologies, and the Politics of Information." *European Journal of International Law* 23, no. 4 (2012): 1089–1123. https://doi.org/10.1093/ejil/chs073.

Altemeyer-Bartscher, Martin, Oliver Holtemöller, Axel Lindner, Andreas Schmalzbauer, and Götz Zeddies. "On the Distribution of Refugees in the EU." *Intereconomics: Review of European Economic Policy* 51 (August 2016): 220–28. http://doi.org/10.1007/s10272-016-0606-y.

Alves, Jaime Amparo. *The Anti-Black City: Police Terror and Black Urban Life in Brazil*. Minneapolis: University of Minnesota Press, 2018.

Amin, Samir. "Accumulation and Development: A Theoretical Model." *Review of African Political Economy* 1, no. 1 (August 1974): 9–26. https://doi.org/10.1080/03056247408703234.

Amnesty International. "Automated Apartheid: How Facial Recognition Fragments, Segregates and Controls Palestinians in the OPT." May 2, 2023. www.amnesty.org/en/documents/mde15/6701/2023/en.

———. "The Human Cost of Fortress Europe: Human Rights Violations against Migrants and Refugees at Europe's Borders." July 2014. https://doi.org/10.1163/2210-7975_HRD-9211-2014033.

———. "Israeli Authorities Using Facial Recognition to Entrench Apartheid." May 2, 2023. www.amnesty.org/en/latest/news/2023/05/israel-opt-israeli-authorities-are-using-facial-recognition-technology-to-entrench-apartheid.

———. "A Licence to Discriminate: Trump's Muslim and Refugee Ban." October 6, 2020. www.amnesty.org.uk/licence-discriminate-trumps-muslim-refugee-ban.

Amoore, Louise. "Algorithmic War: Everyday Geographies of the War on Terror." *Antipode* 41, no. 1 (January 2009): 49–69. https://doi.org/10.1111/j.1467-8330.2008.00655.x.

———. "Biometric Borders: Governing Mobilities in the War on Terror." *Political Geography* 25, no. 3 (March 2006): 336–51. https://doi.org/10.1016/j.polgeo.2006.02.001.

Anderson, Elijah. "The Cosmopolitan Canopy." *The* annals *of the American Academy of Political and Social Science* 595, no. 1 (September 2004): 14–31. https://doi.org/10.1177/0002716204266833.

Ángel Díaz. "New York City Police Department Surveillance Technology." Brennan Center for Justice, October 2019. www.brennancenter.org/our-work/research-reports/new-york-city-police-department-surveillance-technology.

Anil, Merih. "No More Foreigners? The Remaking of German Naturalization and Citizenship Law, 1990–2000." *Dialectical Anthropology* 29, no. 3/4 (2005): 453–70.

Anoush Darabi. "Amsterdam and Barcelona Are Handing Citizens Control of Their Data." May 22, 2018. https://apolitical.co/en/solution_article/amsterdam-and-barcelona-are-handing-citizens-control-of-their-data.

Appadurai, Arjun. "Disjuncture and Difference in the Global Cultural Economy." *Theory, Culture and Society* 7, no. 2/3 (June 1990): 295–310. https://doi.org/10.1177/026327690007002017.

———. *Modernity at Large: Cultural Dimensions of Globalization*. Minneapolis: University of Minnesota Press, 1996.

———. "Number in the Colonial Imagination." In *Orientalism and the Postcolonial Predicament: Perspectives on South Asia*, edited by Carol Appadurai Breckenridge and Peter van der Veer,314–40. Philadelphia: University of Pennsylvania Press, 1993.

Aricat, Rajiv George, and Rich Ling. *Mobile Communication and Low-Skilled Migrants' Acculturation to Cosmopolitan Singapore*. Lanham, MD: Lexington Books, 2018.

Atanasoski, Neda, and Kalindi Vora. *Surrogate Humanity: Race, Robots, and the Politics of Technological Futures*. Durham, NC: Duke University Press, 2019.

Atkinson, Paul, Amanda Coffey, Sara Delamont, John Lofland, and Lyn Lofland, eds. *Handbook of Ethnography*. London: Sage, 2001.

Awosanya, Yinka. "NCC Allegedly Tags Google Free Wi-Fi as 'Illegal,' in a Move That Could Be Intended to Protect Local Telcos." *Techpoint.Africa*, March 12, 2019. https://techpoint.africa/2019/03/12/google-free-wi-fi-ncc-big-brother.

Baldassar, Loretta, and Joanne Pyke. "Intra-Diaspora Knowledge Transfer and 'New' Italian Migration." *International Migration* 52, no. 4 (2014): 128–43. https://doi.org/10.1111/imig.12135.

Balzacq, Thierry. "The Policy Tools of Securitization: Information Exchange, EU Foreign and Interior Policies." *JCMS: Journal of Common Market Studies* 46, no. 1 (2007): 75–100. https://doi.org/10.1111/j.1468-5965.2007.00768.x.

Bandung2 Blog. "Fugitive Decolonial Luddism: A Hauntology." December 11, 2019. https://bandung2blog.wordpress.com/2019/12/11/fugitive-decolonial-luddism-a-hauntology.

Bateman-House, Alison, and Amy Fairchild. "Medical Examination of Immigrants at Ellis Island." *AMA Journal of Ethics* 10, no. 4 (April 2008): 235–41. https://doi.org/10.1001/virtualmentor.2008.10.4.mhst1-0804.

Bauder, Harald. "Toward a Critical Geography of the Border: Engaging the Dialectic of Practice and Meaning." *Annals of the Association of American Geographers* 101, no. 5 (2011): 1126–39. https://doi.org/10.1080/00045608.2011.577356.

Bayor, Ronald H. *Encountering Ellis Island: How European Immigrants Entered America*. How Things Worked. Baltimore: Johns Hopkins University Press, 2014.

Benjamin, Ruha. *Race after Technology: Abolitionist Tools for the New Jim Code*. Hoboken, NJ: Wiley, 2019.

Bennett, Christina, Matthew Foley, and Hanna B. Krebs, eds. "Learning from the Past to Shape the Future: Lessons from the History of Humanitarian Action in Africa." Humanitarian Policy Group working paper, London, October 2016. https://odi.cdn.ngo/media/documents/11148.pdf.

Benton, Meghan. "Digital Litter: The Downside of Using Technology to Help Refugees." Migration Policy Institute, June 18, 2019. www.migrationpolicy.org/article/digital-litter-downside-using-technology-help-refugees.

———. "Smart Inclusive Cities: How New Apps, Big Data, and Collaborative Technologies Are Transforming Immigrant Integration." Migration Policy Institute, September 12, 2014. www.migrationpolicy.org/research/smart-inclusive-cities-new-apps-big-data-and-collaborative-technologies.

Benton, Meghan, and Alex Glennie. "Digital Humanitarianism: How Tech Entrepreneurs Are Supporting Refugee Integration." Migration Policy Institute, October 2016. www.migrationpolicy.org/research/digital-humanitarianism-how-tech-entrepreneurs-are-supporting-refugee-integration.

Bernal, Amiel. "The Epistemic Injustice Anthology: A Review of *The Routledge Handbook of Epistemic Injustice*." *Social Epistemology Review and Reply Collective* 6, no. 11 (2017): 1–8. https://social-epistemology.com/wp-content/uploads/2017/10/bernal_review_rhei1.pdf.

Bernd, Candice. "Facial Recognition Technology Is Aiding ICE in Sanctuary Cities." Truthout, August 8, 2019. https://truthout.org/articles/facial-recognition-technology-is-aiding-ice-in-sanctuary-cities.

Bernstein, Hamutal, with Nicole DuBois. "Bringing Evidence to the Refugee Integration Debate." Urban Institute, April 2018. www.urban.org/sites/default/files/publication/97771/bringing_evidence_to_the_refugee_integration_debate_0.pdf.

Betterplace Lab. "Refugee Tech: The Hype and What Happened Next." May 2019. www.betterplace-lab.org/refugee-tech-the-hype-and-what-happened-next.

Bhattacharya, Ananya. "Hacking the Refugee Crisis." *The Verge*, April 17, 2016. www.theverge.com/2016/4/17/11446268/techfugees-refugee-crisis-europe-syria.

Biesecker, Cal. "Elbit America Develops AI-Based Solutions for Border and Other Security Applications," *Defense Daily*, June 17, 2020. www.elbitamerica.com/news/elbit-america-develops-ai-based-solutions-for-border-and-other-security-applications.

Black, Edwin. *IBM and the Holocaust: The Strategic Alliance between Nazi Germany and America's Most Powerful Corporation*. Washington, DC: Dialog Press, 2012.

Blanke, Bernhard, and Randall Smith. *Cities in Transition: New Challenges, New Responsibilities*. New York: St. Martin's Press, 1999.

Bonilla-Silva, Eduardo. *Racism without Racists: Color-Blind Racism and the Persistence of Racial Inequality in America*. 6th edition. Lanham, MD: Rowman and Littlefield, 2022.

Børsen, Tom, and Lars Botin, eds. *What Is Techno-Anthropology?* Series in Transformational Studies 3. Aalborg: Aalborg University Press, 2013.

Boyd, Danah, and Kate Crawford. "Six Provocations for Big Data." A Decade in Internet Time: Symposium on the Dynamics of the Internet and Society, September 21, 2011. http://dx.doi.org/10.2139/ssrn.1926431.

Brady, Kate. "Germany Launches Digital Strategy to Become Artificial Intelligence Leader." *Deutsche Welle*, November 15, 2018. www.dw.com/en/germany-launches-digital-strategy-to-become-artificial-intelligence-leader/a-46298494.

Brady, Kate. "German Right-Wing Populist Movement Put under Surveillance." *Deutsche Welle*, August 12, 2016. www.dw.com/en/german-right-wing-populist-movement-put-under-surveillance/a-19469904.

Brandom, Russell. "Huawei Worked on Facial Recognition System to Surveil Uighurs, New Report Claims." *The Verge*, December 8, 2020. www.theverge.com/2020/12/8/22163499/huawei-uighur-surveillance-facial-recognition-megvii-uyghur.

Brayne, Sarah. *Predict and Surveil: Data, Discretion, and the Future of Policing*. New York: Oxford University Press, 2021.

Brayne, Sarah, and Angèle Christin. "Technologies of Crime Prediction: The Reception of Algorithms in Policing and Criminal Courts." *Social Problems* 68, no. 3 (August 2021): 608–24. https://doi.org/10.1093/socpro/spaa004.

Brenner, Neil, and Christian Schmid. "Planetary Urbanisation." In *Urban Constellations*, edited by Matthew Gandy, 10–13. Berlin: Jovis, 2011.

Brock, André. "Critical Technocultural Discourse Analysis." *New Media and Society* 20, no. 3 (2018): 1012–30. https://doi.org/10.1177/1461444816677532.

———. *Distributed Blackness: African American Cybercultures*. Critical Cultural Communication. New York: NYU Press, 2019.

Broeders, Dennis. "The New Digital Borders of Europe: EU Databases and the Surveillance of Irregular Migrants." *International Sociology* 22, no. 1 (January 2007): 71–92. https://doi.org/10.1177/0268580907070126.

Broeders, Dennis, and James Hampshire. "Dreaming of Seamless Borders: ICTs and the Pre-emptive Governance of Mobility in Europe." *Journal of Ethnic and Migration Studies* 39, no. 8 (2013): 1201–18. https://doi.org/10.1080/1369183X.2013.787512.

Brook, Chris. *Unruly Cities? Order/Disorder.* London: Routledge, 2006.

Broussard, Meredith. *Artificial Unintelligence: How Computers Misunderstand the World.* Cambridge, MA: MIT Press, 2018.

———. *More Than a Glitch: Confronting Race, Gender, and Ability Bias in Tech.* Cambridge, MA: MIT Press, 2023.

Brown, Anastasia, and Todd Scribner. "Unfulfilled Promises, Future Possibilities: The Refugee Resettlement System in the United States." *Journal on Migration and Human Security* 2, no. 2 (2014): 101–20. https://doi.org/10.1177/233150241400200203.

Browne, Simone. *Dark Matters: On the Surveillance of Blackness.* Durham, NC: Duke University Press, 2015.

Buck, Tobias. "German Start-Ups Attract Record Investment in 2017." *Financial Times,* 10 January 2018. www.ft.com/content/0394dafc-f606-11e7-88f7-5465a6ce1a00.

Buckland, Michael. *Information and Society.* Cambridge, MA: MIT Press, 2017.

Budds, Diana. "A New Report Outlines Privacy Risks for the MTA's Contactless Payment System." *Curbed NY,* October 3, 2019. https://ny.curbed.com/2019/10/3/20895736/mta-omny-privacy-surveillance-report.

Buff, Rachel. "Teaching Crown Heights: The Complex Language of Identity." *Shofar* 15, no. 3 (1997): 19–32.

Buolamwini, Joy, and Timnit Gebru. "Gender Shades: Intersectional Accuracy Disparities in Commercial Gender Classification." Proceedings of the 1st Conference on Fairness, Accountability, and Transparency. *PMLR* 81 (2018): 77–91. https://proceedings.mlr.press/v81/buolamwini18a.html.

Butler, Judith. "Performativity, Precarity and Sexual Politics." *AIBR: Revista de Antropología Iberoamericana* 4, no. 3 (2009): i–xiii. https://doi.org/10.11156/aibr.040303e.

Byrnes, Nanette. "Are Machine Learning Algorithms Biased?' *MIT Technology Review,* June 24, 2016. www.technologyreview.com/s/601775/why-we-should-expect-algorithms-to-be-biased.

Campbell, Zach. "Sci-Fi Surveillance: Europe's Secretive Push into Biometric Technology." *The Guardian*, December 10, 2020. www.theguardian.com/world/2020/dec/10/sci-fi-surveillance-europes-secretive-push-into-biometric-technology.

Cardullo, Paolo, Cesare Di Felicaiantonio, and Rob Kitchin, eds. *The Right to the Smart City*. Bingley, UK: Emerald Publishing, 2019.

Castells, Manuel. *The Rise of the Network Society*. 2nd ed. Malden, MA: Wiley-Blackwell, 2010.

Catterall, Bob. "'Planetary' Urbanisation: Insecure Foundations, the Commodification of Knowledge, and Paradigm Shift." *City* 20, no. 1 (2016): 1–9. https://doi.org/10.1080/13604813.2016.1146009.

Chaar-López, Iván. "Sensing Intruders: Race and the Automation of Border Control." *American Quarterly* 71, no. 2 (2019): 495–518. https://doi.org/10.1353/aq.2019.0040.

Chan, Anita. *Networking Peripheries: Technological Futures and the Myth of Digital Universalism*. Cambridge, MA: MIT Press, 2013.

Cheesman, Margie. "Self-Sovereignty for Refugees? The Contested Horizons of Digital Identity." *Geopolitics* 27, no. 1 (2022): 134–59. https://doi.org/10.1080/14650045.2020.1823836.

Chen, Brian X. "Apple Registers Trademark for 'There's an App for That.'" *Wired*, October 11, 2010. www.wired.com/2010/10/app-for-that.

Cheney-Lippold, John. *We Are Data: Algorithms and the Making of Our Digital Selves*. New York: NYU Press, 2017.

Chin, Rita. *The Guest Worker Question in Postwar Germany*. Cambridge: Cambridge University Press, 2007.

Christin, Angèle. "The Ethnographer and the Algorithm: Beyond the Black Box." *Theory and Society* 49, no. 5 (2020): 897–918. https://doi.org/10.1007/s11186-020-09411-3.

Chung, Jen. "The MTA's New OMNY Scanners Have Cameras in Them (But They're Not Watching You, Yet)." *Gothamist*, June 14, 2019. https://gothamist.com/news/the-mtas-new-omny-scanners-have-cameras-in-them-but-theyre-not-watching-you-yet.

Clavell, Gemma Galdon. "The Political Economy of Surveillance in the (Wannabe) Global City." *Surveillance and Society* 8, no. 4 (2011): 523–26. https://doi.org/10.24908/ss.v8i4.4191.

Clifford, James, and George E. Marcus, eds. *Writing Culture: The Poetics and Politics of Ethnography: A School of American Research Advanced Seminar.* Berkeley: University of California Press, 2009.

CNN. "DeBlasio Defends NY Sanctuary City Policies." *State of the Union*, January 29, 2017. www.cnn.com/videos/tv/2017/01/29/sotu-mayor-de-blasio-on-undocumented-drunk-drivers.cnn.

Cohen, Peter, Robert Hahn, Jonathan Hall, Steven Levitt, and Robert Metcalfe. "Using Big Data to Estimate Consumer Surplus: The Case of Uber." Working paper 22627, National Bureau of Economic Research, Cambridge, MA, September 2016. https://doi.org/10.3386/w22627.

Coleman, Simon, and Pauline von Hellermann. *Multi-sited Ethnography: Problems and Possibilities in the Translocation of Research Methods.* London: Routledge, 2012.

Collier, Paul, and Alexander Betts. *Refuge: Rethinking Refugee Policy in a Changing World.* Oxford: Oxford University Press, 2017.

Cone, Jason. "The Promise of Social Media for Humanitarian Action?' *Harvard Program on Humanitarian Policy and Conflict Research*, May 10, 2012. http://hpcrresearch.org/blog/hpcr/2012-05-10/promise-social-media-humanitarian-action.

Cordero-Guzmán, Héctor R., Robert C. Smith, and Ramón Grosfoguel. *Migration, Transnationalization, and Race in a Changing New York.* Philadelphia: Temple University Press, 2001.

Costanza-Chock, Sasha. *Design Justice: Community-Led Practices to Build the Worlds We Need.* Cambridge, MA: MIT Press, 2020.

———. "Design Justice: Towards an Intersectional Feminist Framework for Design Theory and Practice." Proceedings of the Design Research Society, June 3, 2018. https://ssrn.com/abstract=3189696.

Couldry, Nick, and Ulises Ali Mejias. *The Costs of Connection: How Data Is Colonizing Human Life and Appropriating It for Capitalism.* Culture and Economic Life. Stanford, CA: Stanford University Press, 2019.

Cox, Joseph. "How the U.S. Military Buys Location Data from Ordinary Apps." *Vice*, November 2020. www.vice.com/en/article/jgqm5x/us-military-location-data-xmode-locate-x.

Culbertson, Shelly, James Dimarogonas, Katherine Costello, and Serafina Lanna. *Crossing the Digital Divide: Applying Technology to the Global Refugee Crisis.* Los Angeles: Rand Corporation, 2019.

Currier, Cora. "Prosecuting Parents—and Separating Families—Was Meant to Deter Migration, Signed Memo Confirms." *The Intercept*, September 25, 2018. https://theintercept.com/2018/09/25/family-separation-border-crossings-zero-tolerance.

Danewid, Ida. "White Innocence in the Black Mediterranean: Hospitality and the Erasure of History." *Third World Quarterly* 38, no. 7 (2017): 1674–89. https://doi.org/10.1080/01436597.2017.1331123.

Darrow, Jessica. "The (Re)Construction of the U.S. Department of State's Reception and Placement Program by Refugee Resettlement Agencies." *Journal of the Society for Social Work and Research* 6, no. 1 (2015): 91–119. https://doi.org/10.1086/680341.

Dastin, Jeffrey. "Microsoft to Divest AnyVision Stake, End Face Recognition Investing." *Reuters*, March 28, 2020. www.reuters.com/article/us-microsoft-anyvision-idUSKBN21E3BA.

Davies, James, and Dimitrina Spencer, eds. *Emotions in the Field: The Psychology and Anthropology of Fieldwork Experience.* Stanford, CA: Stanford University Press, 2010.

Davis, Angela. *The Prison Industrial Complex.* San Francisco: Ak Press Audio, 2000.

Davis, Mike, and Justin Akers Chacón. *No One Is Illegal: Fighting Racism and State Violence on the U.S.–Mexico Border.* Updated ed. Chicago: Haymarket Books, 2018.

De Genova, Nicholas. "The European Question: Migration, Race, and Postcoloniality in Europe." *Social Text* 34, no. 3 (2016): 75–102. https://doi.org/10.1215/01642472-3607588.

———. "Spectacles of Migrant 'Illegality': The Scene of Exclusion, the Obscene of Inclusion." *Ethnic and Racial Studies* 36, no. 7 (2013): 1180–98. https://doi.org/10.1080/01419870.2013.783710.

Deakin, Mark, and Husam Al Waer. "From Intelligent to Smart Cities." *Intelligent Buildings International* 3, no. 3 (July 2011): 140–52. https://doi.org/10.1080/17508975.2011.586671.

De Laat, Sonya, and Valérie Gorin. "Iconographies of Humanitarian Aid in Africa." In "Learning from the Past to Shape the Future: Lessons from the History of Humanitarian Action in Africa," edited by Christina Bennett, Matthew Foley, and Hanna B. Krebs, Humanitarian Policy Group working paper, London, October 2016. https://odi.cdn.ngo/media/documents/11148.pdf.

Delcker, Janosch. "Germany's Falling behind on Tech, and Merkel Knows It." *Politico*, July 23, 2018. www.politico.eu/article/germany-falling-behind-china-on-tech-innovation-artificial-intelligence-angela-merkel-knows-it.

Deleuze, Gilles. "Postscript on the Societies of Control." *October* 59 (1992): 3–7.

Derviş, Kemal, and Laurence Chandy. "Are Technology and Globalization Destined to Drive up Inequality?' Brookings, October 5, 2016. www.brookings.edu/research/are-technology-and-globalization-destined-to-drive-up-inequality.

Devereaux, Ryan. "How ICE Operations in New York Set the Stage for a Coronavirus Nightmare in Local Jails." *The Intercept*, March 27, 2020. https://theintercept.com/2020/03/27/immigrants-coronavirus-ice-detention-new-york.

Di Maio, Alessandra. "The Mediterranean, or Where Africa Does (Not) Meet Italy: Andrea Segre's *A Sud Di Lampedusa* (2006)." In *The Cinemas of Italian Migration: European and Transatlantic Narratives*, edited by Sabine Schrader and Daniel Winkler, 41–52. Cambridge: Cambridge Scholars Publishing, 2013.

Dijstelbloem, Huub, and Albert Meijer, eds. *Migration and the New Technological Borders in Europe*. Basingstoke: Palgrave Macmillan, 2011.

Dirlik, Arif. "Developmentalism: A Critique." *Interventions* 16, no. 1 (2014): 30–48. https://doi.org/10.1080/1369801X.2012.735807.

Dlamini, S. Nombuso, Barat Wolfe, Uzo Anucha, and Miu Chung Yan. "Engaging the Canadian Diaspora: Youth Social Identities in a Canadian Border City." *McGill Journal of Education* 44, no. 3 (2009): 405–33. https://doi.org/10.7202/039947ar.

Doorn, Niels van. "Platform Labor: On the Gendered and Racialized Exploitation of Low-Income Service Work in the 'On-Demand' Economy." *Information, Communication and Society* 20, no. 6 (2017): 898–914. https://doi.org/10.1080/1369118X.2017.1294194.

Dreby, Joanna. "How Today's Immigration Enforcement Policies Impact Children, Families, and Communities." Center for American Progress, August 2012. www.americanprogress.org/article/how-todays-immigration-enforcement-policies-impact-children-families-and-communities.

Duan, Xiao-dong, Cun-rui Wang, Xiang-dong Liu, Zhi-jie Li, Jun Wu, and Hai-long Zhang. "Expression of Concern: Ethnic Features Extraction and Recognition of Human Faces." In *2010 2nd International Conference on Advanced Computer Control*, 125–30. New York: IEEE, 2010. https://doi.org/10.1109/ICACC.2010.5487194.

Dunlap, Margi, and Nicholas Montalto. "Out of Many, One: A History of the Immigration and Refugee Services of America Network." International Institute, 1998. www.iistl.org/resources.

Dunleavy, Patrick. *Digital Era Governance: IT Corporations, the State, and E-Government*. Oxford: Oxford University Press, 2006.

Easterling, Keller. *Extrastatecraft: The Power of Infrastructure Space*. London: Verso, 2014.

Ebzeev, Boris. "Little Russia Ain't Just Little Russia." *Fathom*, July 28, 2014. www.fathomaway.com/brighton-beach-walking-tour.

Eisenhower, Dwight D. "Military–Industrial Complex Speech." Avalon Project at the Yale Law School, 1961. https://avalon.law.yale.edu/20th_century/eisenhower001.asp.

Eisenzweig, Tal D. "In the Shadow of Child Protective Services: Noncitizen Parents and the Child–Welfare System." *Yale Law Journal*, November 2018. www.yalelawjournal.org/forum/in-the-shadow-of-child-protective-services.

Elish, M. C. "Don't Call AI Magic." *Data and Society: Points*, January 17, 2018. https://medium.com/datasociety-points/dont-call-ai-magic-142da16db408.

Elish, M. C., and Danah Boyd. "Situating Methods in the Magic of Big Data and Artificial Intelligence." *Communication Monographs* 85, no. 1 (2018): 57–80. https://doi.org/10.1080/03637751.2017.1375130.

Emmer, Pascal, Chris Schweidler, Brenda Salas Neves, and Caroline Rivas. "Technologies for Liberation: Toward Abolitionist Futures." Astraea Lesbian Foundation for Justice, 2020. www.astraeafoundation.org/FundAbolitionTech.

England, Marcia R., and Stephanie Simon. "Scary Cities: Urban Geographies of Fear, Difference and Belonging." *Social and Cultural Geography* 11, no. 3 (2010): 201–7. https://doi.org/10.1080/14649361003650722.

Erel, Umut, Karim Murji, and Zaki Nahaboo. "Understanding the Contemporary Race–Migration Nexus." *Ethnic and Racial Studies* 39, no. 8 (2016): 1339–60. https://doi.org/10.1080/01419870.2016.1161808.

Estabrooks, Maurice. *Programmed Capitalism: Computer-Mediated Global Society: Computer-Mediated Global Society*. London: Routledge, 2017.

Eubanks, Virginia. *Automating Inequality: How High-Tech Tools Profile, Police, and Punish the Poor.* New York: St. Martin's Press, 2018.

———. "Double-Bound: Putting the Power Back into Participatory Research." *Frontiers: A Journal of Women Studies* 30, no. 1 (2009): 107–37. https://doi.org/10.1353/fro.0.0023.

European Commission. "The Digital Skills Gap in Europe." 19 October 2017. https://ec.europa.eu/digital-single-market/en/news/digital-skills-gap-europe.

Falguera, Xavier Casademont, Òscar Prieto-Flores, and Jordi Feu Gelis. "Refugee and Romani Immigrant Populations in Barcelona." In *The Oxford Handbook of Migration Crises*, edited by Cecilia Menjívar, Marie Ruiz, and Immanuel Ness. Oxford Handbooks (2019; online ed., Oxford Academic, September 10, 2018). https://doi.org/10.1093/oxfordhb/9780190856908.013.8.

Fanon, Frantz. *The Wretched of the Earth: Frantz Fanon*. Translated by Richard Philcox. 1961. New York: Grove Press, 2004.

Fejerskov, Adam Moe. "The New Technopolitics of Development and the Global South as a Laboratory of Technological Experimentation." *Science, Technology, and Human Values* 42, no. 5 (2017): 947–68. https://doi.org/10.1177/0162243917709934.

Felter, Claire, and James McBride. "How Does the U.S. Refugee System Work?" Council on Foreign Relations. Last updated March 26, 2024. www.cfr.org/backgrounder/how-does-us-refugee-system-work.

Fernback, Jan. "Selling Ourselves? Profitable Surveillance and Online Communities." *Critical Discourse Studies* 4, no. 3 (2007): 311–30. https://doi.org/10.1080/17405900701656908.

Ferrer, Josep-Ramon. "Barcelona's Smart City Vision: An Opportunity for Transformation." *Field Actions Science Reports. The Journal of Field Actions*, special issue 16 (2017): 70–75. https://journals.openedition.org/factsreports/4367.

Fields, Karen E., and Barbara Jeanne Fields. *Racecraft: The Soul of Inequality in American Life*. London: Verso, 2014.

Finck, Michhle, and Sofia Ranchordas. "Sharing and the City." *Vanderbilt Law Review* 49, no. 5 (2016): https://scholarship.law.vanderbilt.edu/vjtl/vol49/iss5/3.

Fitzgerald, David. "Towards a Theoretical Ethnography of Migration." *Qualitative Sociology* 29, no. 1 (2006): 1–24. https://doi.org/10.1007/s11133-005-9005-6.

Fleming, Sean. "The Two Faces of Personhood: Hobbes, Corporate Agency and the Personality of the State." *European Journal of Political Theory* 20, no. 1 (2017): https://doi.org/10.1177/1474885117731941.

Florida, Richard. "Innovation and Economic Segregation Are Two Sides of the Same Coin in the U.S." *Bloomberg*, October 24, 2017. www.citylab.com/equity/2017/10/how-innovation-leads-to-economic-segregation/543759.

Foner, Nancy. "Immigration History and the Remaking of New York." In *New York and Amsterdam: Immigration and the New Urban Landscape*, edited by Nancy Foner, Jan Rath, Jan Willem Duyvendak, and Rogier van Reekum, 29–51. New York: NYU Press, 2014.

Foner, Nancy, ed. *New Immigrants in New York*. New York: Columbia University Press, 1987.

Foner, Nancy, Jan Rath, Jan Willem Duyvendak, and Rogier van Reekum, eds. *New York and Amsterdam: Immigration and the New Urban Landscape*. New York: NYU Press, 2014.

Fong, Benjamin. *Death and Mastery: Psychoanalytic Drive Theory and the Subject of Late Capitalism*. New York: Columbia University Press, 2016.

Foucault, Michel. *"Society Must Be Defended": Lectures at the Collège de France, 1975–1976*. Translated by David Macey. New York: Picador, 1997.

Francois, George, and Chris George. "Introducing Equiano, a Subsea Cable from Portugal to South Africa." *Google Cloud* (blog), June 28, 2019. https://cloud.google.com/blog/products/infrastructure/introducing-equiano-a-subsea-cable-from-portugal-to-south-africa.

Franklin, Seb. *Control: Digitality as Cultural Logic*. Cambridge, MA: MIT Press, 2015.

Frenkel, Sheera. "Microsoft Employees Protest Work with ICE, as Tech Industry Mobilizes over Immigration." *New York Times*, June 20, 2018. www.nytimes.com/2018/06/19/technology/tech-companies-immigration-border.html.

Fricker, Miranda. *Epistemic Injustice: Power and the Ethics of Knowing*. New York: Oxford University Press, 2007.

———. "Epistemic Justice as a Condition of Political Freedom?' *Synthese* 190, no. 7 (2013): 1317–32. https://doi.org/10.1007/s11229-012-0227-3.

Friedman, Uri. "The Mathematical Equations That Could Decide the Fate of Refugees." *The Atlantic*, September 9, 2015. www.theatlantic.com/international/archive/2015/09/formula-european-union-refugee-crisis/404503.

Frontex. "Artificial Intelligence-Based Capabilities for the European Border and Coast Guard." Final report, March 2021. www.frontex.europa.eu/assets/Publications/Research/Frontex_AI_Research_Study_2020_final_report.pdf.

Fuchs, Christian. "Capitalism or Information Society? The Fundamental Question of the Present Structure of Society." *European Journal of Social Theory* 16, no. 4 (2013): 413–34. https://doi.org/10.1177/1368431012461432.

Fussell, Sidney. "NYC Launches Task Force to Study How Government Algorithms Impact Your Life." *Gizmodo*, May 16, 2018. https://gizmodo.com/nyc-launches-task-force-to-study-how-government-algorit-1826087643.

Gandy, Oscar H. *The Panoptic Sort: A Political Economy of Personal Information*. Critical Studies in Communication and in the Cultural Industries. Boulder, CO: Westview, 1993.

Garner, Steve. "The European Union and the Racialization of Immigration, 1985–2006." *Race/Ethnicity: Multidisciplinary Global Contexts* 1, no. 1 (2007): 61–87. https://muse.jhu.edu/pub/3/article/369122.

Georgiou, Myria. "*City of Refuge* or Digital Order? Refugee Recognition and the Digital Governmentality of Migration in the City." *Television and New Media* 20, no. 6 (2019): 600–616. https://doi.org/10.1177/1527476419857683.

———. *Diaspora, Identity, and the Media: Diasporic Transnationalism and Mediated Spatialities*. Urban Communication. Cresskill, NJ: Hampton Press, 2006.

———. "Does the Subaltern Speak? Migrant Voices in Digital Europe." *Popular Communication* 16, no. 1 (2018): 45–57. https://doi.org/10.1080/15405702.2017.1412440.

Geraghty, Lena, Tina Lee, Julia Glickman, and Brooks Rainwater. "Future of Cities: Cities and the Metaverse." National League of Cities, Centre for City Solutions, 2022. www.nlc.org/wp-content/uploads/2022/04/CS-Cities-and-the-Metaverse_v4-Final-1.pdf.

Gesley, Jenny. "Germany: The Development of Migration and Citizenship Law in Postwar Germany." Law Library of Congress, March 2017. https://tile.loc.gov/storage-services/service/ll/llglrd/2016478971/2016478971.pdf.

Gilroy, Paul. *The Black Atlantic: Modernity and Double Consciousness*. Cambridge, MA: Harvard University Press, 2003.

Glaeser, Edward L. "Urban Colossus: Why Is New York America's Largest City?" *FRBNY Economic Policy Review*, December 2005, 7–24. www.newyorkfed.org/medialibrary/media/research/epr/05v11n2/0512glae.pdf.

Glendinning, Chellis. "Notes toward a Neo-Luddite Manifesto." Anarchist Library, 1990. https://theanarchistlibrary.org/library/chellis-glendinning-notes-toward-a-neo-luddite-manifesto.

Goodfellow, Maya. *Hostile Environment: How Immigrants Became Scapegoats*. London: Verso, 2019.

Grabner-Kräuter, Sonja, and Sofie Bitter. "Trust in Online Social Networks: A Multifaceted Perspective." *Forum for Social Economics* 44, no. 1 (2015): 48–68. https://doi.org/10.1080/07360932.2013.781517.

Graham, Thomas. "Barcelona Is Leading the Fightback against Smart City Surveillance." *Wired*, May 18, 2018. www.wired.co.uk/article/barcelona-decidim-ada-colau-francesca-bria-decode.

Granqvist, Manne. "Assessing ICT in Development: A Critical Perspective." In *Media and Glocal Change: Rethinking Communication for Development*, edited by Oscar Hemer and Thomas Tufte, 285–96. Gothenburg: Nordicom, 2005.

Green, Ben. *The Smart Enough City: Putting Technology in Its Place to Reclaim Our Urban Future*. Strong Ideas. Cambridge, MA: MIT Press, 2019.

Greenberg, Zoe. "The IDs Were Meant to Protect Immigrants. Are They a Liability?' *New York Times*, July 10, 2018. www.nytimes.com/2018/07/10/nyregion/idnyc-fort-drum-silva-barrios.html.

Guibernau, Montserrat. "Migration and the Rise of the Radical Right." Policy Network, March 2010. https://esquerraeuropea.wordpress.com/wp-content/uploads/2012/02/migrationandradicalright.pdf.

Gutierrez Rodriguez, Encarnacion. "The Coloniality of Migration and the 'Refugee Crisis': On the Asylum–Migration Nexus, the Transatlantic White European Settler Colonialism–Migration and Racial Capitalism." *Refuge* 34 (2018): 16–28. https://doi.org/10.7202/1050851ar.

Haddad, Angela T., and Richard Senter. "The Relationship of Technology to Workers' Alienation." *Sociological Focus* 50, no. 2 (2017): 159–82. https://doi.org/10.1080/00380237.2017.1251755.

Hall, Suzanne M. "Migrant Margins: The Streetlife of Discrimination." *Sociological Review* 66, no. 5 (2018): 968–83. https://doi.org/10.1177/0038026118771282.

———. "Migrant Urbanisms: Ordinary Cities and Everyday Resistance." *Sociology* 49, no. 5 (2015): 853–69. https://doi.org/10.1177/0038038515586680.

———. "Mooring 'Super-Diversity' to a Brutal Migration Milieu." *Ethnic and Racial Studies* 40, no. 9 (1 2017): 1562–73. https://doi.org/10.1080/01419870.2017.1300296.

———. "The Politics of Belonging." *Identities* 20, no. 1 (2013): 46–53. https://doi.org/10.1080/1070289X.2012.752371.

———. "Super-Diverse Street: A 'Trans-Ethnography' across Migrant Localities." *Ethnic and Racial Studies* 38, no. 1 (2015): 22–37. https://doi.org/10.1080/01419870.2013.858175.

Hamel, Pierre, Henri Lustiger-Thaler, and Margit Mayer. *Urban Movements in a Globalising World*. London: Routledge, 2003.

Hamid, Eba, and Josh Williams. "New York City's Newest Immigrant Enclaves." *New York Times*, June 8, 2013. www.nytimes.com/interactive/2013/06/09/nyregion/new-york-citys-newest-immigrant-enclaves.html.

Hamilton, Amber M. "A Genealogy of Critical Race and Digital Studies: Past, Present, and Future." *Sociology of Race and Ethnicity* 6, no. 3 (2020): 292–301. https://doi.org/10.1177/2332649220922577.

Hancox, Dan. "Is This the World's Most Radical Mayor?" *The Guardian*, May 26, 2016. www.theguardian.com/world/2016/may/26/ada-colau-barcelona-most-radical-mayor-in-the-world.

Harvey, David. "The 'New' Imperialism: Accumulation by Dispossession." In *Karl Marx*, edited by Bertell Ollman and Kevin B. Anderson, 213–37. London: Routledge, 2017.

Hasselskog, Malin, and Isabell Schierenbeck. "The Ownership Paradox: Continuity and Change." *Forum for Development Studies* 44, no. 3 (2017): 323–33. https://doi.org/10.1080/08039410.2017.1384530.

Healy, Joshua, Daniel Nicholson, and Andreas Pekarek. "Should We Take the Gig Economy Seriously?" *Labour and Industry: A Journal of the Social and Economic Relations of Work* 27, no. 3 (2017): 232–48. https://doi.org/10.1080/10301763.2017.1377048.

Heckmann, Friedrich. "Understanding the Creation of Public Consensus: Migration and Integration in Germany." Migration Policy Institute, June

2016. www.migrationpolicy.org/research/understanding-creation-public-consensus-migration-and-integration-germany-2005-2015.

Heilbroner, Robert. "Technology and Capitalism." *Social Research* 64, no. 3 (1997): 1321–25.

Heng, Geraldine. *The Invention of Race in the European Middle Ages*. Cambridge: Cambridge University Press, 2018.

———. "The Invention of Race in the European Middle Ages I: Race Studies, Modernity, and the Middle Ages1: Invention of Race in the European Middle Ages." *Literature Compass* 8, no. 5 (May 2011): 258–74. https://doi.org/10.1111/j.1741-4113.2011.00790.x.

Hernæs, Christoffer O. "Is Technology Contributing to Increased Inequality?" *TechCrunch*, March 29, 2017. https://techcrunch.com/2017/03/29/is-technology-contributing-to-increased-inequality.

Hessy, Elliott. "A Home Away from Home: Innovations to Support Refugee Inclusion in Cities." *Digital Leaders*, January 22, 2020. https://digileaders.com/a-home-away-from-home-innovations-to-support-refugee-inclusion-in-cities.

Hill, Kashmir. "The Violent Opt-Out: The Neo-Luddites Attacking Drones and Google Glass." *Forbes*, July 16, 2014. www.forbes.com/sites/kashmirhill/2014/07/15/the-violent-opt-out-people-destroying-drones-and-google-glass.

Hindman, Matthew Scott. *The Myth of Digital Democracy*. Princeton, NJ: Princeton University Press, 2009.

Hirsch, Shirin. "Racism, 'Second Generation' Refugees and the Asylum System." *Identities* 26, no. 1 (2019): 88–106. https://doi.org/10.1080/1070289X.2017.1361263.

Hoffmann, Anna Lauren, Nicholas Proferes, and Michael Zimmer. "'Making the World More Open and Connected': Mark Zuckerberg and the Discursive Construction of Facebook and Its Users." *New Media and Society* 20, no. 1 (2016): 199–218. https://doi.org/10.1177/1461444816660784.

Hong, Sun-ha. *Technologies of Speculation: The Limits of Knowledge in a Data-Driven Society*. New York: NYU Press, 2020.

Houtum, Henk Van, and Roos Pijpers. "The European Union as a Gated Community: The Two-Faced Border and Immigration Regime of the EU." *Antipode* 39, no. 2 (2007): 291–309. https://doi.org/10.1111/j.1467-8330.2007.00522.x.

Human Rights Watch. "China: Police 'Big Data' Systems Violate Privacy, Target Dissent." November 29, 2017. www.hrw.org/news/2017/11/19/china-police-big-data-systems-violate-privacy-target-dissent.

Hunter, Walt. "The Story behind the Poem on the Statue of Liberty." *The Atlantic*, January 16, 2018. www.theatlantic.com/entertainment/archive/2018/01/the-story-behind-the-poem-on-the-statue-of-liberty/550553.

Ilse, Jess. "Prince of Wales Becomes First UK Patron of International Rescue Committee." *Royal Central*, January 12, 2020. https://royalcentral.co.uk/uk/prince-of-wales-becomes-first-uk-patron-of-international-rescue-committee-136142.

———. "Royal Germany Tour Recap: The Prince of Wales and the Duchess of Cornwall Split Their Time in Berlin and Munich." *Royal Central*, May 10, 2019. https://royalcentral.co.uk/uk/royal-germany-tour-recap-the-prince-of-wales-and-the-duchess-of-cornwall-split-their-time-in-berlin-and-munich-125343.

International Rescue Committee. "The Prince of Wales and the Duchess of Cornwall Meet Refugee Women in Berlin." May 9, 2019. www.rescue-uk.org/press-release/prince-wales-and-duchess-cornwall-meet-refugee-women-berlin.

Jacob, David. "Workeer: A Job Board for Refugees." *edenspiekermann_*, October 1, 2015. www.edenspiekermann.com/magazine/workeer-a-job-board-for-refugees.

Jacobsen, Katja Lindskov. "Experimentation in Humanitarian Locations: UNHCR and Biometric Registration of Afghan Refugees." *Security Dialogue* 46, no. 2 (2015): 144–64. https://doi.org/10.1177/0967010614552545.

Jaeggi, Rahel. *Alienation*. New York: Columbia University Press, 2014.

Jasanoff, Sheila, and Sang-Hyun Kim. *Dreamscapes of Modernity: Sociotechnical Imaginaries and the Fabrication of Power.* Chicago: University of Chicago Press, 2015.

Jensen, Eric Allen, and Alexander Charles Laurie. *Doing Real Research: A Practical Guide to Social Research*. Los Angeles: Sage, 2016.

Johnson, Gaye Theresa, and Alex Lubin, eds. *Futures of Black Radicalism*. New York: Verso, 2017.

Jones, Hannah, and Emma Jackson, eds. *Stories of Cosmopolitan Belonging: Emotion and Location*. London: Taylor and Francis, 2014.

Kasinitz, Philip. *Caribbean New York: Black Immigrants and the Politics of Race.* Anthropology of Contemporary Issues. Ithaca, NY: Cornell University Press, 1992.

Kayyali, Dia. "EFF Files Amicus Brief in Case That Seeks to Hold IBM Responsible for Facilitating Apartheid in South Africa." Electronic Frontier Foundation, February 5, 2015. www.eff.org/deeplinks/2015/02/eff-files-amicus-brief-case-seeks-hold-ibm-responsible-facilitating-apartheid.

Keil, Roger. "The Empty Shell of the Planetary: Re-rooting the Urban in the Experience of the Urbanites." *Urban Geography* 39, no. 10 (2018): 1589–1602. https://doi.org/10.1080/02723638.2018.1451018.

Kelley, Robin D. G. "What Did Cedric Robinson Mean by Racial Capitalism?' *Boston Review*, January 12, 2017. http://bostonreview.net/race/robin-d-g-kelley-what-did-cedric-robinson-mean-racial-capitalism.

Kelly, Nicola. "Facial Recognition Smartwatches to Be Used to Monitor Foreign Offenders in UK." *The Guardian*, August 5, 2022. www.theguardian.com/politics/2022/aug/05/facial-recognition-smartwatches-to-be-used-to-monitor-foreign-offenders-in-uk.

Kingston, Lindsey N. "Biometric Identification, Displacement, and Protection Gaps." In *Digital Lifeline? ICTs for Refugees and Displaced Persons*, edited by Maitland, Carleen, 35–54. Cambridge, MA: MIT Press, 2018.

Kirk, Jeffrey. *10 Million to 1: Refugee Resettlement: A How-To Guide.* Bloomington, IN: Balboa Press, 2011.

Kofman, Ava. "Are New York's Free LinkNYC Internet Kiosks Tracking Your Movements?" *The Intercept*, September 8, 2018. https://theintercept.com/2018/09/08/linknyc-free-wifi-kiosks.

Kusmer, Ken. "Ind. Cancels IBM's $1.3 Billion Welfare Contract." *San Diego Union-Tribune*, October 15, 2009. www.sandiegouniontribune.com/sdut-us-indiana-welfare-101509-2009oct15-story.html.

Lagemann, Ellen Condliffe. *The Politics of Knowledge: The Carnegie Corporation, Philanthropy, and Public Policy.* Chicago: University of Chicago Press, 1992.

Lamdan, Sarah. *Data Cartels: The Companies That Control and Monopolize Our Information.* Stanford, CA: Stanford University Press, 2023.

Lampen, Claire. "Yes, LinkNYC Kiosks Are Giant Data-Harvesting Surveillance Cameras, Obviously." *Gothamist*, April 25, 2019. https://gothamist.com/news/yes-linknyc-kiosks-are-giant-data-harvesting-surveillance-cameras-obviously.

Lane, Jeffery. *The Digital Street*. Oxford: Oxford University Press, 2018.

Latonero, Mark, Keith Hiatt, Antonella Napolitano, Giulia Clericetti, and Melanie Penagos. "Digital Identity in the Migration and Refugee Context: Italy Case Study." Data and Society Research Institute, April 15, 2019. https://apo.org.au/node/231471.

Latonero, Mark, and Paula Kift. "On Digital Passages and Borders: Refugees and the New Infrastructure for Movement and Control." *Social Media and Society* 4, no. 1 (2018): https://doi.org/10.1177/2056305118764432.

Latour, Bruno. "On Interobjectivity." *Mind, Culture, and Activity* 3, no. 4 (1996): 228–45. https://doi.org/10.1207/s15327884mca0304_2.

Lawson, Clive. "An Ontology of Technology: Artefacts, Relations and Functions." *Techné: Research in Philosophy and Technology* 12, no. 1 (2008): 48–64. https://doi.org/10.5840/techne200812114.

Lazarus, Emma. "The New Colossus." In *The Norton Introduction to Literature*, edited by Kelly J. Mays, shorter 14th ed., 752. New York: W. W. Norton, 2022.

Lecher, Colin. "Thousands of ICE Employees Can Access License Plate Reader Data, Emails Show." *The Verge*, March 13, 2019. www.theverge.com/2019/3/13/18262141/ice-license-plate-reader-database-aclu-emails.

Lefebvre, Henri. *The Production of Space*. Translated by Donald Nicholson-Smith. Nachdr. Malden, MA: Blackwell, 2011.

———. *Writings on Cities*. Edited and translated by Eleonore Kofman, and Elizabeth Lebas. Cambridge, MA: Blackwell, 1996.

Lewicki, Aleksandra. "Race, Islamophobia and the Politics of Citizenship in Post-Unification Germany." *Patterns of Prejudice* 52, no. 5 (2018): 496–512. https://doi.org/10.1080/0031322X.2018.1502236.

Llamas-Rodriguez, Juan. *Border Tunnels: A Media Theory of the U.S.–Mexico Underground*. Minneapolis: University of Minnesota Press, 2023.

Lobo, Arun Peter, and Joseph J. Salvo. "The Newest New Yorker, 2013 Edition." NYC Planning, 2013. https://www1.nyc.gov/assets/planning/download/pdf/planning-level/nyc-population/nny2013/chapter3.pdf.

Loukissas, Yanni A. *All Data Are Local: Thinking Critically in a Data-Driven Society*. Cambridge, MA: MIT Press, 2019.

Lyon, David. "Surveillance, Snowden, and Big Data: Capacities, Consequences, Critique." *Big Data and Society* 1, no. 2 (2014): https://doi.org/10.1177/2053951714541861.

Lyons, Kim. "The Consortium behind LinkNYC Kiosks Is 'Delinquent' and Owes the City Millions." *The Verge*, March 5, 2020. www.theverge.com/2020/3/5/21166057/linknyc-wifi-free-kiosk-google-new-york-sidewalk-labs-payments-revenue.

Maass, Dave. "Four Steps Facebook Should Take to Counter Police Sock Puppets." Electronic Frontier Foundation, April 14, 2019. www.eff.org/deeplinks/2019/04/facebook-must-take-these-four-steps-counter-police-sock-puppets.

Mac, Ryan, Caroline Haskins, and Logan McDonald. "Clearview AI's Facial Recognition Tech Is Being Used by the Justice Department, ICE, and the FBI." *Buzzfeed News*, February 27, 2020. www.buzzfeednews.com/article/ryanmac/clearview-ai-fbi-ice-global-law-enforcement.

Madianou, Mirca. "Humanitarian Campaigns in Social Media." *Journalism Studies* 14, no. 2 (2013): 249–66. https://doi.org/10.1080/1461670X.2012.718558.

———. "Technocolonialism: Digital Innovation and Data Practices in the Humanitarian Response to Refugee Crises." *Social Media and Society* 5, no. 3 (2019): https://doi.org/10.1177/2056305119863146.

Maeckelbergh, Marianne. "From Digital Tools to Political Infrastructure." In *The Sage Handbook of Resistance*, edited by David Courpasson and Steven Vallas, 280–96. London: Sage, 2016.

———. "Mobilizing to Stay Put: Housing Struggles in New York City: Mobilizing to Stay Put in New York." *International Journal of Urban and Regional Research* 36, no. 4 (2012): 655–73.

Mahmoudi, Matthew. "Digital Borders in [Smart] Cities of Refuge." Open Society Foundations, 2020. [Internal report, not available online.]

Maitland, Carleen, ed. *Digital Lifeline? ICTs for Refugees and Displaced Persons*. Cambridge, MA: MIT Press, 2018.

Manhattan Times News. "City Unveils New IDNYC Features." January 16, 2020. www.manhattantimesnews.com/city-unveils-new-idnyc-featuresciudad-presenta-nuevas-caracteristicas-de-la-idnyc.

Mann, Steve, Jason Nolan, and Barry Wellman. "Sousveillance: Inventing and Using Wearable Computing Devices for Data Collection in Surveillance Environments." *Surveillance and Society* 1, no. 3 (2003): 331–55. https://doi.org/10.24908/ss.v1i3.3344.

March, Hug, and Ramon Ribera-Fumaz. "Smart Contradictions: The Politics of Making Barcelona a Self-Sufficient City." *European Urban and Regional Studies* 23, no. 4 (2016): 816–30. https://doi.org/10.1177/0969776414554488.

Marcuse, Peter. "The Enclave, the Citadel, and the Ghetto: What Has Changed in the Post-Fordist U.S. City." *Urban Affairs Review* 33, no. 2 (1997): 228–64. https://doi.org/10.1177/107808749703300206.

Markel, Howard, and Alexandra Minna Stern. "The Foreignness of Germs: The Persistent Association of Immigrants and Disease in American Society." *Milbank Quarterly* 80, no. 4 (December 2002): 757–88. https://doi.org/10.1111/1468-0009.00030.

Marosi, Richard. "Hundreds of Migrant Children Have Been Sent to New York. Here's How They Spend Their Days." *Los Angeles Times*, July 9, 2018. www.latimes.com/nation/immigration/la-me-migrant-children-new-york-20180709-story.html.

Martin, Bonnie. "Slavery's Invisible Engine: Mortgaging Human Property." *Journal of Southern History* 76, no. 4 (2010): 817–66. https://doi.org/10.1080/0144039X.2017.1397334.

Martone, Eric. "Treacherous 'Saracens' and Integrated Muslims: The Islamic Outlaw in Robin Hood's Band and the Re-Imagining of English Identity, 1800 to the Present." *Miscelánea: A Journal of English and American Studies* 40 (2009): 53–76. https://dialnet.unirioja.es/servlet/articulo?codigo=3150101.

Mathers, Andrew, and Mario Novelli. "Researching Resistance to Neoliberal Globalization: Engaged Ethnography as Solidarity and Praxis." *Globalizations* 4, no. 2 (2007): 229–49. https://doi.org/10.1080/14747730701345259.

May, Stephen A. "Critical Ethnography." In *Encyclopedia of Language and Education: Research Methods in Language and Education*, edited by Nancy H. Hornberger and David Corson, 197–206. Encyclopedia of Language and Education. Dordrecht: Springer Netherlands, 1997.

Maynard, Robyn. "Black Life and Death across the U.S.–Canada Border: Border Violence, Black Fugitive Belonging, and a Turtle Island View of Black Liberation." *Critical Ethnic Studies* 5, no. 1/2 (2019): 124–51. https://doi.org/10.5749/jcritethnstud.5.1-2.0124.

Mbembe, Achille. "Bodies as Borders." *From the European South* 4 (2019): 5–18. http://europeansouth.postcolonialitalia.it/journal/2019-4/2.Mbembe.pdf.

———. "The Idea of a Borderless World." *Africa Is a Country*, November 11, 2018. https://africasacountry.com/2018/11/the-idea-of-a-borderless-world.

McCarroll, Estefania. "Weapons of Mass Deportation: Big Data and Automated Decision-Making Systems in Immigration Law." *Georgetown Immigration Law Journal* 34 (2020): 705–31. www.law.georgetown.edu/immigration-law-journal/in-print/volume-34-number-3-spring-2020/weapons-of-mass-deportation-big-data-and-automated-decision-making-systems-in-immigration-law.

McCarthy, Craig. "NYPD Issues Policy on Facial Recognition after Nearly Decade of Use." *New York Post*, March 12, 2020. https://nypost.com/2020/03/12/nypd-issues-policy-on-facial-recognition-software-after-nearly-a-decade-of-use.

McFarlane, Colin. "The Entrepreneurial Slum: Civil Society, Mobility and the Co-production of Urban Development." *Urban Studies* 49, no. 13 (2012): 2795–2816. https://doi.org/10.1177/0042098012452460.

McFarlane, Colin, and Ola Söderström. "On Alternative Smart Cities." *City* 21, no. 3/4 (2017): 312–28. https://doi.org/10.1080/13604813.2017.1327166.

McIlwain, Charlton D. *Black Software: The Internet and Racial Justice, from the AfroNet to Black Lives Matter.* New York: Oxford University Press, 2020.

McKenzie, Robert L. "Refugees Don't Just Come to Nations; They Move to Cities." Brookings, October 3, 2016. www.brookings.edu/blog/metropolitan-revolution/2016/10/03/refugees-dont-just-come-to-nations-they-move-to-cities.

McLaughlin, Eugene, and John Muncie. *The Sage Dictionary of Criminology.* London: Sage, 2019.

McLean, Heather. "In Praise of Chaotic Research Pathways: A Feminist Response to Planetary Urbanization." *Environment and Planning D: Society and Space* 36, no. 3 (2017): https://doi.org/10.1177/0263775817713751.

Meaker, Morgan. "Europe Is Using Smartphone Data as a Weapon to Deport Refugees." *Wired*, July 2, 2018. www.wired.co.uk/article/europe-immigration-refugees-smartphone-metadata-deportations.

Meier, Patrick. *Digital Humanitarians: How Big Data Is Changing the Face of Humanitarian Response.* Boca Raton, FL: CRC Press, 2015.

———. "New Information Technologies and Their Impact on the Humanitarian Sector." *International Review of the Red Cross* 93, no. 884 (2011): 1239–63. https://doi.org/10.1017/S1816383112000318.

Meijer, Albert. "Digitization and Political Accountability in the USA and the Netherlands: Convergence or Reproduction of Differences?' *Electronic*

Journal of E-Government 5, no. 2 (2007): 213–24. https://academic-publishing.org/index.php/ejeg/article/view/478.

Mertia, Sandeep. "FCJ-217 Socio-Technical Imaginaries of a Data-Driven City: Ethnographic Vignettes from Delhi." *Fibreculture Journal*, no. 29 (July 2017): https://doi.org/10.15307/fcj.29.217.2017.

Mezzadra, Sandro. "Abolitionist Vistas of the Human: Border Struggles, Migration and Freedom of Movement." *Citizenship Studies* 24, no. 4 (2020): 424–40. https://doi.org/10.1080/13621025.2020.1755156.

Mignolo, Walter D. "Introduction." *Cultural Studies* 21, no. 2/3 (2007): 155–67. https://doi.org/10.1080/09502380601162498.

Migracode. "Migracode Barcelona: A Background Article." January 8, 2020. https://migracode.openculturalcenter.org/a-background-article.

Miguel, Cristina. "Visual Intimacy on Social Media: From Selfies to the Co-construction of Intimacies through Shared Pictures." *Social Media and Society* 2, no. 2 (2016): https://doi.org/10.1177/2056305116641.7.

Mijente. "Palantir's Technology Used in Mississippi Raids Where 680 Were Arrested." October 4, 2019. www.ohchr.org/sites/default/files/Documents/Issues/Racism/SR/RaceBordersDigitalTechnologies/Palantirs_technology_used_in_Mississippi_raids_where_680_were_arrested.pdf.

Mijente, Immigrant Defense Project, and National Immigration Project. "Who's behind ICE? The Tech and Data Companies Fueling Deportations." 2018. https://mijente.net/wp-content/uploads/2018/10/WHO%E2%80%99S-BEHIND-ICE_-The-Tech-and-Data-Companies-Fueling-Deportations-_v1.pdf.

Miller, Ben. "Cubic Wins $500M Contract to Overhaul NYC Subway and Bus Fare System." *Government Technology*, October 26, 2017. www.govtech.com/biz/cubic-wins-500m-contract-to-overhaul-nyc-subway-and-bus-fare-system.html.

Miller, Todd. "More Than a Wall: Corporate Profiteering and the Militarization of US Borders." Transnational Institute, September 2019. www.tni.org/en/publication/more-than-a-wall-0.

Millington, Gareth. "The Cosmopolitan Contradictions of Planetary Urbanization: The Cosmopolitan Contradictions of Planetary Urbanization." *British Journal of Sociology* 67, no. 3 (2016): 476–96. https://doi.org/10.1111/1468-4446.12200.

———. *Urbanization and the Migrant in British Cinema: Spectres of the City*. London: Palgrave Macmillan, 2016.

Miroff, Nick, Amy Goldstein, and Maria Sacchetti. "'Deleted' Families: What Went Wrong with Trump's Family-Separation Effort." *Washington Post*, July 2018. www.washingtonpost.com/local/social-issues/deleted-families-what-went-wrong-with-trumps-family-separation-effort/2018/07/28/54bcdcc6-90cb-11e8-8322-b5482bf5e0f5_story.html.

Mirzoeff, Nicholas. "Artificial Vision, White Space and Racial Surveillance Capitalism." *AI and Society* 36 (2020): 1295–1305. https://doi.org/10.1007/s00146-020-01095-8.

Mischke, Juditch. "Germany Passes Controversial Migration Law." *Politico*, June 7, 2019. www.politico.eu/article/germany-passes-controversial-migration-law.

Mohler, Jeremy. "Just Do It: The Manufacturing of the Neoliberal Subject." *GNOVIS* 1 (Fall 2015): https://web.archive.org/web/20201001074329/www.gnovisjournal.org/2015/12/04/just-do-it-the-manufacturing-of-the-neoliberal-subject.

Molnar, Petra. "Technological Testing Grounds: Migration Management Experiments and Reflections from the Ground Up." EDRi, November 2020. https://edri.org/wp-content/uploads/2020/11/Technological-Testing-Grounds.pdf.

Monshipouri, Mahmood. *Information Politics, Protests, and Human Rights in the Digital Age*. Cambridge: Cambridge University Press, 2016.

Morozov, Evgeny. "Digital Intermediation of Everything: At the Intersection of Politics, Technology and Finance." Council of Europe, n.d. https://rm.coe.int/digital-intermediation-of-everything-atthe-intersection-of-politics-t/168075baba.

———. *To Save Everything, Click Here: Technology, Solutionism, and the Urge to Fix Problems That Don't Exist*. London: Penguin, 2013.

Morozov, Evgeny, Richard Barbrook, and Francesca Bria. "Digital Democracy and Technological Sovereignty." *openDemocracy*, December 13, 2016. www.opendemocracy.net/en/digitaliberties/digital-democracy-and-technological-sovereignty.

Mossberger, Karen, Caroline J. Tolbert, and Ramona S. McNeal. *Digital Citizenship: The Internet, Society, and Participation*. Cambridge, MA: MIT Press, 2007.

Mozilla, Bryan Pon, Caribou Digital, and Mark Surman. "An Ad-Supported Internet Isn't Going to Be Sustainable in Emerging Markets." *Quartz Africa*, August 1, 2017. https://qz.com/africa/1042994/an-ad-supported-internet-isnt-going-to-be-sustainable-in-emerging-markets.

Mozur, Paul. "One Month, 500,000 Face Scans: How China Is Using A.I. to Profile a Minority." *New York Times*, April 14, 2019. www.nytimes.com/2019/04/14/technology/china-surveillance-artificial-intelligence-racial-profiling.html.

Muggah, Robert. "Safe Havens: Why Cities Are Crucial to the Global Refugee Crisis." World Economic Forum, June 2017. www.weforum.org/agenda/2017/06/safe-havens-why-cities-are-crucial-to-the-global-refugee-crisis.

Muggah, Robert, and Adriana Erthal Abdenur. "Refugees and the City: The Twenty-First-Century Front Line." World Refugee Council Research Paper 2, World Refugee Council, July 2018. www.cigionline.org/publications/refugees-and-city-twenty-first-century-front-line.

Muller, Benjamin J. "(Dis)Qualified Bodies: Securitization, Citizenship and 'Identity Management.'" *Citizenship Studies* 8, no. 3 (2010): 279–94. https://doi.org/10.1080/1362102042000257005.

Muñiz, Ana. *Borderland Circuitry: Immigration Surveillance in the United States and Beyond*. Oakland: University of California Press, 2022.

Munshi, Kaivan. "Networks in the Modern Economy: Mexican Migrants in the U.S. Labor Market." *Quarterly Journal of Economics* 118, no. 2 (2003): 549–99. https://doi.org/10.1162/003355303321675455.

Murray, John. "Barcelona: Europe's Smart City." Medium, March 19, 2019. https://medium.com/primalbase/barcelona-europes-smart-city-9a2f169ecbc0.

Murthy, Dhiraj. "Digital Ethnography: An Examination of the Use of New Technologies for Social Research." *Sociology* 42, no. 5 (2008): 837–55. https://doi.org/10.1177/0038038508094565.

Muselli, Laure, Mathieu O'Neil, Fred Pailler, and Stefano Zacchiroli. "The Open-Source World Is More and More Closed." *Le Monde diplomatique*, January 2022. https://mondediplo.com/2022/01/09digital.

Navarro-Remesal, Víctor, and Beatriz Pérez Zapata. "Who Made Your Phone? Compassion and the Voice of the Oppressed in *Phone Story* and *Burn the Boards*." *Open Library of Humanities* 4, no. 1 (2018): https://doi.org/10.16995/olh.209.

Nelson, Alondra. *The Social Life of DNA: Race, Reparations, and Reconciliation after the Genome*. Boston: Beacon Press, 2016.

Nevins, Joseph. *Dying To Live: A Story of U.S. Immigration in an Age of Global Apartheid*. San Francisco: City Lights, 2008.

Nevins, Joseph, and Allan Nevins. *Operation Gatekeeper: The Rise of the "Illegal Alien" and the Making of the U.S.–Mexico Boundary*. New York: Psychology Press, 2002.

New York Civil Liberties Union. "Automatic License Plate Readers." July 23, 2015. www.nyclu.org/en/automatic-license-plate-readers.

———. "Testimony of the New York Civil Liberties Union in Opposition to a Smart Chip Being Added to IDNYC." October 8, 2019. www.nyclu.org/resources/policy/testimonies/testimony-new-york-civil-liberties-union-opposition-smart-chip-being-added-idnyc.

New Economy Project. "Letter to Mayor Bill de Blasio on Proposal to Add Financial Technology to IDNYC Cards." 12 September 2019. www.neweconomynyc.org/resource/letter-to-mayor-bill-de-blasio-on-proposal-to-add-financial-technology-to-idnyc-cards.

Nkonde, Mutale. "Automated Anti-Blackness: Facial Recognition in Brooklyn, New York." *Harvard Kennedy School Journal of African American Policy*, 2019. https://pacscenter.stanford.edu/wp-content/uploads/2020/12/mutalenkonde.pdf.

Noble, Safiya Umoja. *Algorithms of Oppression: How Search Engines Reinforce Racism*. New York: NYU Press, 2018.

Noestlinger, Nette, Matthias Baehr, and Elizabeth Howcroft. "Worldcoin Says Will Allow Companies, Governments to Use Its ID System." *Reuters*, August 2, 2023. www.reuters.com/technology/worldcoin-says-will-allow-companies-governments-use-its-id-system-2023-08-02.

NYC.gov. "De Blasio Administration Announces NYCx Technology Leadership Advisory Council Members." Official website of the City of New York, January 11, 2018. www.nyc.gov/office-of-the-mayor/news/027-18/de-blasio-administration-nycx-technology-leadership-advisory-council-members.

———. "De Blasio Administration Announces Winner of Competition to Replace Payphones." Official website of the City of New York, November 17, 2014. http://www1.nyc.gov/office-of-the-mayor/news/923–14/de-blasio-administration-winner-competition-replace-payphones-five-borough.

———. "Mayor's Office of Immigrant Affairs Releases Third Annual Report, Solidifying NYC's Commitment to Protect the Rights of All New Yorkers,

Regardless of Immigration Status." Official website of the City of New York, April 14, 2020. www.nyc.gov/site/immigrants/about/press-releases/20200414-moia-releases-third-annual-report.page.

———. "On Citizenship Day, de Blasio Administration Launches New Campaign to Promote Naturalization." Official website of the City of New York, September 17, 2018. https://www1.nyc.gov/site/immigrants/about/press-releases/09-17-2018.page.

NYU Press. "An Interview with John Cheney-Lippold, Author of *We Are Data*." *From The Square* (blog), March 3, 2017. www.fromthesquare.org/we-are-data.

Ojo, Tokunbo. "Neo-Gramscian Approach and Geopolitics of ICT4D Agenda." *Global Media Journal, Canadian Edition* 9, no. 1 (2016): 23–35. https://epe.lac-bac.gc.ca/100/201/300/global_media_journal/v09n01/www.gmj.uottawa.ca/1601/v9i1_ojo_abstract.html.

Olivares, José. "Are "Sanctuary Cities" Doing Enough? A New Report Shows How to Really Fight Trump's Deportation Machine." *The Intercept*, March 21, 2018. https://theintercept.com/2018/03/21/sanctuary-cities-deportation-ice-trump.

Olwig, Karen Fog. "'Transnational' Socio-cultural Systems and Ethnographic Research: Views from an Extended Field Site." *International Migration Review* 37, no. 3 (2003): 787–811. https://doi.org/10.1111/j.1747–7379.2003.tb00158.x.

O'Malley, James. "Stop Acting Surprised That Refugees Have Smartphones." *The Independent*, September 7, 2015. www.independent.co.uk/voices/comment/surprised-syrian-refugees-have-smartphones-well-sorry-break-you-you-re-idiot-10489719.html.

———. "Surprised That Syrian Refugees Have Smartphones? Sorry to Break This to You, but You're an Idiot." *The Independent*, September 7, 2015. www.independent.co.uk/voices/comment/surprised-syrian-refugees-have-smartphones-well-sorry-break-you-you-re-idiot-10489719.html.

Omi, Michael, and Howard Winant. *Racial Formation in the United States: From the 1960s to the 1990s*. 2nd ed. New York: Routledge, 1994.

O'Neil, Cathy. *Weapons of Math Destruction: How Big Data Increases Inequality and Threatens Democracy*. London: Penguin, 2016.

O'Neill, Maggie. *Asylum, Migration and Community*. Bristol: Policy Press, 2010.

One Young World. "How This Entrepreneur Is Using Blockchain for Humanitarian Projects." Medium, February 23, 2018. www.medium.com/@OneYoungWorld_/how-this-entrepreneur-is-using-blockchain-for-humanitarian-projects-b3d744c50ec5.

OpenAI. "Introducing ChatGPT." November 30, 2022. https://openai.com/blog/chatgpt.

Orwell, George. *1984*. 1949. London: Arcturus, 2021.

Paasi, Anssi. "Border Studies Reanimated: Going beyond the Territorial/Relational Divide." *Environment and Planning A: Economy and Space* 44, no. 10 (2012): 2303–9. https://doi.org/10.1068/a45282.

Pasquale, Frank. *The Black Box Society: The Secret Algorithms That Control Money and Information*. Cambridge, MA: Harvard University Press, 2015.

Pegg, David. "UK Awards Border Contract to Firm Criticised over Role in US Deportations." *The Guardian*, September 17, 2020. www.theguardian.com/politics/2020/sep/17/uk-awards-border-contract-to-firm-criticised-over-role-in-us-deportations.

Penney, Jonathon W. "Chilling Effects: Online Surveillance and Wikipedia Use." *Berkeley Technology Law Journal* 31, no. 1 (2016): 117–82. http://dx.doi.org/10.15779/Z38SS13.

Penny, Laurie. "Laurie Penny on a Tale of Two Cities: How San Francisco's Tech Boom Is Widening the Gap between Rich and Poor." *New Statesman*, April9,2014.www.newstatesman.com/laurie-penny/2014/04/tale-two-cities-how-san-franciscos-tech-boom-widening-gap-between-rich-and-poor.

Perlman, J. E. "The Metamorphosis of Marginality: Four Generations in the Favelas of Rio de Janeiro." *The* annals *of the American Academy of Political and SocialScience*606,no.1(2006):154–77.https://doi.org/10.1177/0002716206288826.

Perritt, Henry H. "The Internet as a Threat to Sovereignty? Thoughts on the Internet's Role in Strengthening National and Global Governance." *Indiana Journal of Global Legal Studies* 5, no. 2 (1998): 423–42. www.repository.law.indiana.edu/ijgls/vol5/iss2/4.

Peterson, Becky. "New York Beat Out San Francisco as a Venture Capital Powerhouse." *Business Insider Nederland*, October 26, 2017. www.businessinsider.nl/new-york-beat-out-san-francisco-as-a-venture-capital-powerhouse-2017-10.

Pierce, Sarah, and Jessica Bolter. "Dismantling and Reconstructing the U.S. Immigration System: A Catalog of Changes under the Trump Presidency."

Migration Policy Institute, July 2020. www.migrationpolicy.org/research/us-immigration-system-changes-trump-presidency.

Pinto, Nick. "Google Is Transforming NYC's Payphones into a 'Personalized Propaganda Engine.'" *Village Voice*, July 6, 2016. www.villagevoice.com/2016/07/06/google-is-transforming-nycs-payphones-into-a-personalized-propaganda-engine.

Plitt, Amy. "How to Apply for NYC's Affordable Housing Lottery." *Curbed NY*, May 17, 2017. https://ny.curbed.com/2017/5/17/15649294/new-york-city-housing-lottery-affordable-apartments.

———. "Q&A: Dan Doctoroff on 'Greater Than Ever' and Rebuilding New York City." *Curbed NY*, September 27, 2017. https://ny.curbed.com/2017/9/27/16369024/dan-doctoroff-bloomberg-administration-interview.

Ponzanesi, Sandra. "Digital Diasporas: Postcoloniality, Media and Affect." *Interventions* 22, no. 8 (2020): 977–93. https://doi.org/10.1080/1369801X.2020.1718537.

Price, Marie, and Lisa Benton-Short. "Counting Immigrants in Cities across the Globe." Migration Policy Institute, January 1, 2007. www.migrationpolicy.org/article/counting-immigrants-cities-across-globe.

———. "Immigrants and World Cities: From the Hyper-Diverse to the Bypassed." *GeoJournal* 68, no. 2 (2007): 103–17. https://doi.org/10.1007/s10708-007-9076-x.

Prins, C., D. Broeders, H. Griffioen, A.-G. Keizer, and E. Keymolen. *iGovernment*. Amsterdam: Amsterdam University Press, 2011.

Privacy International. "All Roads Lead to Palantir." October 29, 2020. https://privacyinternational.org/report/4271/all-roads-lead-palantir.

Pruchniewska, Urszula. "'A Group That's Just Women for Women': Feminist Affordances of Private Facebook Groups for Professionals." *New Media and Society* 21, no. 6 (2019): 1362–79. https://doi.org/10.1177/1461444818822490.

Pyysiäinen, Jarkko, Darren Halpin, and Andrew Guilfoyle. "Neoliberal Governance and 'Responsibilization' of Agents: Reassessing the Mechanisms of Responsibility-Shift in Neoliberal Discursive Environments." *Distinktion: Journal of Social Theory* 18, no. 2 (2017): 215–35. https://doi.org/10.1080/1600910X.2017.1331858.

Quijano, Anibal. "Coloniality of Power, Eurocentrism, and Latin America." *Nepantla: Views from South* 1, no. 3 (2000): 533–80. https://muse.jhu.edu/article/23906.

Qurashi, Fahid. "The Prevent Strategy and the UK 'War on Terror': Embedding Infrastructures of Surveillance in Muslim Communities." *Palgrave*

Communications 4, no. 1 (2018): 1–13. www.nature.com/articles/s41599-017-0061-9.

Rajaram, Prem Kumar. "Refugees as Surplus Population: Race, Migration and Capitalist Value Regimes." *New Political Economy* 23, no. 5 (2018): 627–39. https://doi.org/10.1080/13563467.2017.1417372.

Reuters. "Germany Steps up Warnings about Right-Wing Identitarian Movement." 11 July 2019. www.reuters.com/article/us-germany-farright-idUSKCN1U61G1.

Reynolds, Matt. "Citizens Give up Data in Blockchain Project to Improve Cities." *New Scientist*, May 22, 2017. www.newscientist.com/article/2131950-citizens-give-up-data-in-blockchain-project-to-improve-cities.

Rhoades, Robert E. "Foreign Labor and German Industrial Capitalism, 1871–1978: The Evolution of a Migratory System." *American Ethnologist* 5, no. 3 (1978): 553–73. https://doi.org/10.1525/ae.1978.5.3.02a00080.

Richardson, Allissa V. *Bearing Witness While Black: African Americans, Smartphones, and the New Protest #Journalism*. New York: Oxford University Press, 2019.

Rickards, Lauren, Brendan Gleeson, Mark Boyle, and Cian O'Callaghan. "Urban Studies after the Age of the City." *Urban Studies* 53, no. 8 (2016): 1523–41. https://doi.org/10.1177/0042098016640640.

Rist, Ray C. "Migration and Marginality: Guestworkers in Germany and France." *Daedalus* 108, no. 2 (1979): 95–108.

Robbins, Liz. "De Blasio Defends New York Policies on Immigration." *New York Times*, June 28, 2017. www.nytimes.com/2017/06/28/nyregion/bill-de-blasio-defends-new-york-policies-on-immigration.html.

———. "He Delivered Pizza to an Army Base in Brooklyn. Now He Faces Deportation." *New York Times*, June 6, 2018. www.nytimes.com/2018/06/06/nyregion/pizza-delivery-ice-deportation.html.

Robinson, Cedric J. *Black Marxism: The Making of the Black Radical Tradition*. Chapel Hill: University of North Carolina Press, 1983.

Rostow, W. W. "The Stages of Economic Growth." *Economic History Review* 12, no. 1 (1959): 1–16. https://doi.org/10.2307/2591077.

Rubinstein, Dana, and Joe Anuta. "City Hall Calls Google-Backed LinkNYC Consortium 'Delinquent.'" *Politico*, March 3, 2020. www.politico.com/states/new-york/albany/story/2020/03/03/city-hall-calls-google-backed-linknyc-consortium-delinquent-1264966.

Ruddick, Sue, Linda Peake, Gökbörü S Tanyildiz, and Darren Patrick. "Planetary Urbanization: An Urban Theory for Our Time?' *Environment and PlanningD:SocietyandSpace*36,no.3(2017):https://doi.org/10.1177/0263775817721489.

Rush, Nayla. "'Private' Refugee Resettlement Agencies Mostly Funded by the Government." Center for Immigration Studies, August 10, 2018. https://cis.org/Rush/Private-Refugee-Resettlement-Agencies-Mostly-Funded-Government.

Sadowski, Jathan. "Google Wants to Run Cities without Being Elected: Don't Let It." *The Guardian*, October 24, 2017. www.theguardian.com/commentisfree/2017/oct/24/google-alphabet-sidewalk-labs-toronto.

Salter, Mark B. "Theory of the / : The Suture and Critical Border Studies." *Geopolitics* 17, no. 4 (2012): 734–55. https://doi.org/10.1080/14650045.2012.660580.

Sánchez Nicolás, Elena. "Digital Gap: 42% of EU Citizens Lack Basic Digital Skills." *EUobserver*, June 11, 2020. https://euobserver.com/social/148629.

Sanders, Anna. "Council Bill May Block Smart Chips in New York Municipal IDs." *Government Technology*, September 2019. www.govtech.com/policy/Council-Bill-May-Block-Smart-Chips-in-New-York-Municipal-IDs.html.

Sanyal, Romola. "Urbanizing Refuge: Interrogating Spaces of Displacement." *International Journal of Urban and Regional Research* 38, no. 2 (2014): 558–72. https://doi.org/10.1111/1468-2427.12020.

Schliess, Gero. "Where Have Berlin's Refugees Gone?" *Deutsche Welle*, May 24, 2018. www.dw.com/en/where-have-berlins-refugees-gone/a-19395764.

Schmitt, Paul, Daniel Iland, Elizabeth Belding, and Mariya Zheleva. "Cellular and Internet Connectivity for Displaced Populations." In *Digital Lifeline? ICTs for Refugees and Displaced Persons*, edited by Maitland, Carleen, 115–36. Cambridge, MA: MIT Press, 2018.

Scholz, Trebor. *Uberworked and Underpaid: How Workers Are Disrupting the Digital Economy*. Cambridge: Polity Press, 2017.

Schrader, Stuart. *Badges without Borders: How Global Counterinsurgency Transformed American Policing*. American Crossroads 56. Oakland: University of California Press, 2019.

Semple, Kirk. "Immigration Remakes and Sustains New York, Report Finds." *New York Times*, December 18, 2013. www.nytimes.com/2013/12/19

/nyregion/chinese-diaspora-transforms-new-yorks-immigrant-population-report-finds.html.

Shapiro, Aaron. *Design, Control, Predict: Logistical Governance in the Smart City*. Minneapolis: University of Minnesota Press, 2020.

Shepherd, Laura J., and Caitlin Hamilton. *Understanding Popular Culture and World Politics in the Digital Age*. London: Routledge, 2016.

Shoichet, Catherine E. "The Death Toll in ICE Custody Is the Highest It's Been in 15 Years." *CNN*, September 30, 2020. www.cnn.com/2020/09/30/us/ice-deaths-detention-2020/index.html.

Sicherman, Barbara, and Carol Hurd Green. *Notable American Women: The Modern Period: A Biographical Dictionary*. Cambridge, MA: Harvard University Press, 1980.

Slisco, Aila. "What Is Palantir? Never Profitable Company, Born with Help of CIA Seed Money, Makes Market Debut." *Newsweek*, October 1, 2020. www.newsweek.com/what-palantir-never-profitable-company-born-help-cia-seed-money-makes-market-debut-1535509.

Smith, Helena. "Shocking Images of Drowned Syrian Boy Show Tragic Plight of Refugees." *The Guardian*, September 2, 2015. www.theguardian.com/world/2015/sep/02/shocking-image-of-drowned-syrian-boy-shows-tragic-plight-of-refugees.

Smith, Rich. "Cubic Wins $19.9 Million Israeli Defense Forces Contract." *Motley Fool*, May 10, 2013. www.fool.com/investing/general/2013/05/10/cubic-wins-199-million-israeli-defense-forces-cont.aspx.

Smythe, S. A. "The Black Mediterranean and the Politics of Imagination." *Middle East Report* 286 (Spring 2018): 3–9.

Sowell, Thomas. *Say's Law: An Historical Analysis*. Princeton, NJ: Princeton University Press, 1972.

Spectrum News. "NYPD Seeks Woman Seen Smashing OMNY Screens with Hammer." February 7, 2020. www.ny1.com/nyc/all-boroughs/news/2020/02/07/mta-day-of-action-omny-screen-smash.

Srinivasan, Ramesh, and Ajit Pyati. "Diasporic Information Environments: Reframing Immigrant-Focused Information Research." *Journal of the American Society for Information Science and Technology* 58, no. 12 (2007): 1734–44. https://doi.org/10.1002/asi.20658.

Standing, Guy. *The Precariat: The New Dangerous Class*. Rev. New York: Bloomsbury, 2016.

Stanfill, Mel. "The Interface as Discourse: The Production of Norms through Web Design." *New Media and Society* 17, no. 7 (2015): 1059–74. https://doi.org/10.1177/1461444814520873.

Stein, Amelia. "Keller Easterling: Playing Spaces." *Guernica*, May 15, 2015. www.guernicamag.com/playing-spaces.

Stenum, Helle. "The Body-Border: Governing Irregular Migration through Biometric Technology." *Spheres*, June 27, 2017. http://spheres-journal.org/the-body-border-governing-irregular-migration-through-biometric-technology.

Stoler, Ann Laura, ed. *Imperial Debris: On Ruins and Ruination*. Durham, NC: Duke University Press, 2013.

Stolton, Samuel. "Innovation Funding and Digital Skills Crisis: More Can Be Done, Says EIT Digital CEO." *Euractiv*, September 13, 2018. www.euractiv.com/section/future-connectivity/news/innovation-funding-and-digital-skills-crisis-more-can-be-done-says-eit-chief.

Strickland, Patrick. "Syrian Refugees in Athens Hunger Strike for Relocation." *Al Jazeera*, November 14, 2017. www.aljazeera.com/news/2017/11/syrian-refugees-athens-hunger-strike-relocation-171113112516541.html.

Suarez-Villa, Luis. "The E-Economy and the Rise of Technocapitalism: Networks, Firms, and Transportation." *Growth and Change* 34, no. 4 (2003): 390–414. https://doi.org/10.1046/j.0017-4815.2003.00227.x.

———. *Technocapitalism: A Critical Perspective on Technological Innovation and Corporatism*. Philadelphia: Temple University Press, 2012.

Surveillance Technologies Oversight Project. "OMNY Surveillance Oh My: New York City's Expanding Transit Surveillance Apparatus." October 1, 2019. www.stopspying.org/omny.

Swanson, Carl. "Surveying Post-Bloomberg New York with Sidewalk Labs CEO Dan Doctoroff." *Intelligencer*, August 23, 2017. https://nymag.com/intelligencer/2017/08/surveying-new-york-with-sidewalk-labs-ceo-dan-doctoroff.html.

Taylor, Linnet, Gargi Sharma, Aaron Martin, and Shazade Jameson, eds. *Data Justice and COVID-19: Global Perspectives*. London: Meatspace Press, 2020.

Tazzioli, Martina. "Containment through Mobility: Migrants' Spatial Disobediences and the Reshaping of Control through the Hotspot System." *Journal of Ethnic and Migration Studies* 44, no. 16 (2018): 2764–79. https://doi.org/10.1080/1369183X.2017.1401514.

———. "How Cashless Programmes to Support Refugees' Independence Can Restrict Their Freedoms." *The Conversation*, May 8, 2019. http://theconversation.com/how-cashless-programmes-to-support-refugees-independence-can-restrict-their-freedoms-113623.

Tazzioli, Martina, and Glenda Garelli. "Containment beyond Detention: The Hotspot System and Disrupted Migration Movements across Europe." *Environment and Planning D: Society and Space* 38, no. 6 (2020): 1009–27. https://doi.org/10.1177/0263775818759335.

Terranova, Tiziana. *Network Culture: Politics for the Information Age.* London: Pluto, 2004.

Thatcher, Jim, David O'Sullivan, and Dillon Mahmoudi. "Data Colonialism through Accumulation by Dispossession: New Metaphors for Daily Data." *Environment and Planning D: Society and Space* 34, no. 6 (2016): 990–1006. https://doi.org/10.1177/0263775816633195.

Thomas, Elise. "Tagged, Tracked and in Danger: How the Rohingya Got Caught in the UN's Risky Biometric Database." *Wired*, March 12, 2018. www.wired.co.uk/article/united-nations-refugees-biometric-database-rohingya-myanmar-bangladesh.

Thompson, Carol B. "Philanthrocapitalism: Appropriation of Africa's Genetic Wealth." *Review of African Political Economy* 41, no. 141 (2014): 389–405. https://doi.org/10.1080/03056244.2014.901946.

Thomsen, Jacqueline. "Comey Quotes Statue of Liberty Poem in Response to Trump 'S—hole' Remark." *The Hill*, January 12, 2018. https://thehill.com/blogs/blog-briefing-room/news/368675-comey-quotes-statue-of-liberty-poem-in-response-to-trump.

Tilley, Helen. *Africa as a Living Laboratory: Empire, Development, and the Problem of Scientific Knowledge, 1870–1950.* Chicago: University of Chicago Press, 2011.

Todorov, Tzvetan. "'Race,' Writing, and Culture." Translated by Loulou Mack. *Critical Inquiry* 13, no. 1 (1986): 171–81. https://doi.org/10.1086/448380.

Toyama, Kentaro. *Geek Heresy: Rescuing Social Change from the Cult of Technology.* New York: PublicAffairs, 2015.

Trueman, Matt. "A Light That Never Goes Out: Luddite Rebellion Returns to Manchester." *The Guardian*, July 15, 2019. www.theguardian.com/stage/2019

/jul/15/kandinsky-theatre-luddism-there-is-a-light-that-never-goes-out-luddite-rebellion.

Tsibolane, Pitso, and Irwin Brown. "Principles for Conducting Critical Research Using Postcolonial Theory in ICT4D Studies. *GlobDev 2016* 3 (2016): https://aisel.aisnet.org/globdev2016/3.

Tsing, Anna Lowenhaupt. *The Mushroom at the End of the World: On the Possibility of Life in Capitalist Ruins*. Princeton, NJ: Princeton University Press, 2015.

Tulumello, Simone. "From 'Spaces of Fear' to 'Fearscapes': Mapping for Reframing Theories about the Spatialization of Fear in Urban Space." *Space and Culture*, June 24, 2015. https://doi.org/10.1177/1206331215579716.

Turk, Victoria. "The UK's Tech Sector Faces a Tougher Talent Battle Post-Brexit." *Wired UK*, November 30, 2017. www.wired.co.uk/article/uk-tech-faces-tougher-talent-battle-post-brexit.

Turkle, Sherry. *Alone Together: Why We Expect More from Technology and Less from Each Other*. New Yor: Basic Books, 2011.

Turß, Daniela. "Invading Refugees' Phones: Digital Forms of Migration Control in Germany and Europe." Gesellschaft für Freiheitsrechte, February 2020.https://freiheitsrechte.org/uploads/publications/Digital/Study_Invading-Refugees-Phones_Digital-Forms-of-Migration-Control-Gesellschaft_fuer_Freiheitsrechte_2019.pdf.

UNHCR. "Connecting Refugees: How Internet and Mobile Connectivity Can Improve Refugee Well-Being and Transform Humanitarian Action." June 2016. www.unhcr.org/innovation/wp-content/uploads/2018/02/20160707-Connecting-Refugees-Web_with-signature.pdf.

UNHCR Innovation. "Chasing Opportunity: How New Approaches Can Pave the Way for Refugee Self-Reliance." March 9, 2016. www.unhcr.org/innovation/chasing-opportunity-how-new-approaches-can-pave-the-way-for-refugee-self-reliance.

Valencia, Adrián Sotelo. *The Future of Work*. Leiden: Brill, 2015.

Vaughan-Williams, Nick. "The UK Border Security Continuum: Virtual Biopolitics and the Simulation of the Sovereign Ban." *Environment and Planning D: Society and Space* 28, no. 6 (December 2010): 1071–83. https://doi.org/10.1068/d13908.

Vaughan-Williams, Nick, and Maria Pisani. "Migrating Borders, Bordering Lives: Everyday Geographies of Ontological Security and Insecurity in

Malta." *Social and Cultural Geography* 21, no. 5 (2020): 651–73. https://doi.org/10.1080/14649365.2018.1497193.

Vigo, Julian. *Earthquake in Haiti: The Pornography of Poverty and the Politics of Development*. London: Baobab Tree Books, 2015.

Villa-Nicholas, Melissa. *Data Borders: How Silicon Valley Is Building an Industry around Immigrants*. Oakland: University of California Press, 2023.

Vincent, James. "Forty Percent of 'AI Startups' in Europe Don't Actually Use AI, Claims Report." *The Verge*, March 5, 2019. www.theverge.com/2019/3/5/18251326/ai-startups-europe-fake-40-percent-mmc-report.

Wacquant, Loïc. *Urban Outcast: A Comparative Sociology of Advanced Marginality*. Malden, MA: Polity Press, 2008.

Walia, Harsha. *Undoing Border Imperialism*. Oakland, CA: AK Press, 2014.

Wallerstein, Immanuel Maurice. "The Modern World-System as a Capitalist World-Economy." In *World-Systems Analysis: An Introduction*, 23–41. Durham, NC: Duke University Press, 2004.

———. *World-Systems Analysis: An Introduction*. Durham, NC: Duke University Press, 2004.

Wallis, Emma. "'Not Good for the Economy': German Official Warns against Long-Term Poverty for Refugees and Asylum Seekers." *InfoMigrants*, November 27, 2019. www.infomigrants.net/en/post/21141/not-good-for-the-economy-german-official-warns-against-long-term-poverty-for-refugees-and-asylum-seekers.

Walsham, Geoff. "ICT4D Research: Reflections on History and Future Agenda." *Information Technology for Development* 23, no. 1 (2017): 18–41. https://doi.org/10.1080/02681102.2016.1246406.

Walters, Donna K. H. "IBM to End Its Presence in S. Africa." *Los Angeles Times*, October 22, 1986. www.latimes.com/archives/la-xpm-1986-10-22-mn-6806-story.html.

Wang, Tricia. "Why Big Data Needs Thick Data." Medium, January 20, 2016. https://medium.com/ethnography-matters/why-big-data-needs-thick-data-b4b3e75e3d7.

Wasik, Zosia. "Migrant Crisis Triggers a Wave of Tech Innovation." *Financial Times*, October 26, 2017. www.ft.com/content/e53197ee-8904-11e7-afd2-74b8ecd34d3b.

Weber, Beverly. "The German Refugee 'Crisis' after Cologne: The Race of Refugee Rights." *English Language Notes* 54, no. 2 (2016): 77–92. https://muse.jhu.edu/article/711445.

Wei, Zhao, and Michael A. Peters. "'Intelligent Capitalism' and the Disappearance of Labour: Whitherto Education?' *Educational Philosophy and Theory* 51, no. 8 (2019): 757–66. https://doi.org/10.1080/00131857.2018.1519775.

Weissert, Will, and Zeke Miller. "Mexico Agrees to Invest $1.5B in 'Smart' Border Technology." *AP News*, July 12, 2022. https://apnews.com/article/russia-ukraine-biden-immigration-climate-and-environment-120f8a3fc440e3b2ccc-ce6100e65b912.

Weitzberg, Keren. "Machine-Readable Refugees." *LRB Blog*, September 14, 2020. www.lrb.co.uk/blog/2020/september/machine-readable-refugees.

Wiggers, Kyle. "LinkNYC's 5 Million Users Make 500,000 Phone Calls Each Month." *VentureBeat*, September 29, 2018. https://venturebeat.com/2018/09/29/linknycs-gigabit-kiosks-hit-1-billion-sessions-and-5-million-users.

———. "LinkNYC's 6 Million Users Have Used 8.6 Terabytes of Data." *VentureBeat*, March 28, 2019. https://venturebeat.com/2019/03/28/linknycs-6-million-users-have-downloaded-8-6-terabytes-of-data.

Wilding, Raelene. "Transnational Ethnographies and Anthropological Imaginings of Migrancy." *Journal of Ethnic and Migration Studies* 33, no. 2 (2007): 331–48. https://doi.org/10.1080/13691830601154310.

Williams, James. *Stand Out of Our Light: Freedom and Resistance in the Attention Economy*. New York: Cambridge University Press, 2018.

Willis, Katharine S., and Alessandro Aurigi. *Digital and Smart Cities*. London: Routledge, 2017.

Woodman, Spencer. "Palantir Provides the Engine for Donald Trump's Deportation Machine." *The Intercept*, March 2, 2017. https://theintercept.com/2017/03/02/palantir-provides-the-engine-for-donald-trumps-deportation-machine.

Worldcoin. "Worldcoin Begins Rollout of 1.5k Orbs to Meet Global Demand for World ID." July 24, 2023. https://worldcoin.org/blog/announcements/worldcoin-begins-rollout-orbs-meet-global-demand-world-id.

Zaborowski, Rafal, and Myria Georgiou. "Gamers versus Zombies? Visual Mediation of the Citizen/Non-Citizen Encounter in Europe's 'Refugee

Crisis.'" *Popular Communication* 17, no. 2 (2019): 92–108. https://doi.org/10.1080/15405702.2019.1572150.

Zuboff, Shoshana. *The Age of Surveillance Capitalism: The Fight for a Human Future at the New Frontier of Power.* New York: PublicAffairs, 2018.

———. "Surveillance Capitalism and the Challenge of Collective Action." *New Labor Forum* 28, no. 1 (2019): 10–29. https://doi.org/10.1177/1095796018819461.

Zucker, Norman L. "Refugee Resettlement in the United States: The Role of the Voluntary Agencies." *Michigan Journal of International Law* 3, no. 1 (1982): 155–77. https://repository.law.umich.edu/mjil/vol3/iss1/8.

Zukin, Sharon. *The Innovation Complex: Cities, Tech, and the New Economy.* Oxford: Oxford University Press, 2020.

Index

Founded in 1893,
UNIVERSITY OF CALIFORNIA PRESS
publishes bold, progressive books and journals on topics in the arts, humanities, social sciences, and natural sciences—with a focus on social justice issues—that inspire thought and action among readers worldwide.

The UC PRESS FOUNDATION
raises funds to uphold the press's vital role as an independent, nonprofit publisher, and receives philanthropic support from a wide range of individuals and institutions—and from committed readers like you. To learn more, visit ucpress.edu/supportus.

www.ingramcontent.com/pod-product-compliance
Lightning Source LLC
Chambersburg PA
CBHW020845270326
41928CB00006B/549

9780520397019